CHEMICALS & SOCIETY

a guide to the new chemical age

HUGH D. CRONE

Head, Personnel Protection Group

Materials Research Laboratories, Melbourne, Australia

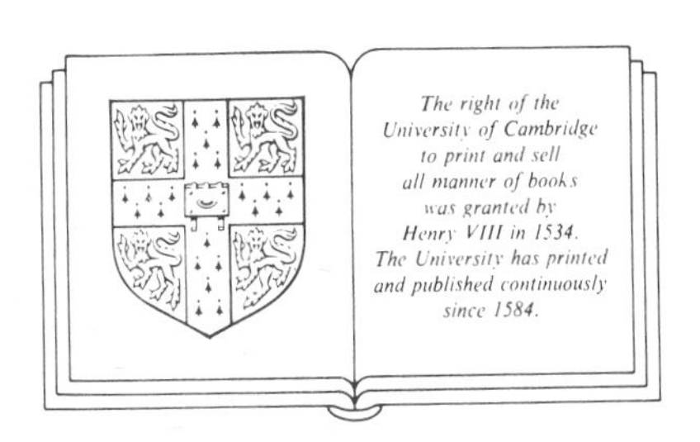

CAMBRIDGE UNIVERSITY PRESS

Cambridge

New York Port Chester

Melbourne Sydney

Published by the Press Syndicate of the University of Cambridge
The Pitt Building, Trumpington Street, Cambridge CB2 1 RP
40 West 20th Street, New York, NY 10011, USA
10 Stamford Road, Oakleigh, Melbourne 3166, Australia

First published 1986
Reprinted 1987, 1988, 1990

Printed in Great Britain at the University Press, Cambridge

British Library cataloguing in publication data

Crone, Hugh D.
Chemicals and society: a guide to the
new chemical age.
1. Chemicals – Social aspects
I. Title
306'.45 QD 39.2

Library of Congress cataloguing in publication data

Crone, Hugh D.
Chemicals and society
Includes index.
1. Chemical – Environmental aspects. 2. Pollution – Environmental aspects. I. Title
TD195.C45C76 1986 363.1'79 85-29934

ISBN 0 521 30869 0 hardback
ISBN 0 521 31359 7 paperback

Contents

Preface

This book has to a large degree written itself. I have found myself over the past few years having to present technical data about chemicals to educated persons who have no technical background. Therefore I have had to think about many of the current concerns about chemicals, then translate the information into plain terms. Thus the raw material for the book was created and, further, I could see the need for wider dissemination of the type of information I was producing. It was evident that intelligent people had problems in assessing chemical matters because they did not have the conceptual background, which in itself is relatively simple. The reasons for this lack are two-fold. Firstly, the necessary background is not presented in popular media, or in non-specialist education systems; bear in mind that I am referring to the concepts necessary to digest the mass of information which is presented, not that information itself. Secondly, the plethora of chemical fact and fancy with which the public is bombarded originates from special interest groups whose aims are to mould your opinion, not to guide you to a logical conclusion. My aim in presenting this book is therefore to remedy this situation, and give you some tools with which to appraise the Chemical World in rational terms.

Expressions of opinion in this book are entirely my own, but of course I owe much to various persons who have been my mentors over the years; Bob Bridges, Tom Keen and Shirley Freeman. The latter, especially, has never despaired of trying to educate me during many tea breaks. Many others have also helped me in various ways. Rita Tkaczyk and Rene Higgins both have the uncanny ability to read my writing and type from it. I am also grateful to my family in allowing me time to think in the intermissions between cello, violin and piano.

Let me once again assure you, the reader, that I do not intend to storm your mind with the janissaries of prejudice, self-interest and obscurantism, but to tempt you with reason. There is little in this world to suggest we are in the Age of Reason, no more than was revealed to Candide in a previous era of supposed light. Nevertheless, we can try.

Melbourne, December 1985 Hugh Crone

Units

Système International (SI) units are used; the abbreviations and the full names are set out below. Also given are the magnitudes of the units relative to that most commonly used. This relative magnitude is given as a power of ten, and is also set out longhand as series of zeros in order to emphasise the magnitude of the differences.

Abbreviation	Full name	Relative magnitude	
TIME			
s	second		1
m	minute		60
h	hour		3600
MASS			
pg	picogram	10^{-12} g	0.000 000 000 001
ng	nanogram	10^{-9} g	0.000 000 001
μg	microgram	10^{-6} g	0.000 001
mg	milligram	10^{-3} g	0.001
g	gram	1 g	1
kg	kilogram	10^{3} g	1000
t	tonne	10^{6} g	1000 000
VOLUME			
l	litre	1 l	1
m^3	cubic metre	10^{3} l	1000
LENGTH			
μm	micrometre	10^{-6} m	0.000 001
mm	millimetre	10^{-3} m	0.001
m	metre	1 m	1
km	kilometre	10^{3} m	1000
CONCENTRATION			
ppb	parts per billion	1 part in	1 000 000 000 parts
ppm	parts per million	1 part in	1 000 000 parts

Notes on concentration units

1. The US convention for a billion is used, i.e. a thousand million, rather than the British meaning of a million million.
2. The term parts per million may be used on a weight basis when analysing liquids or solids, when 1 ppm approximately equals 1 mg/l. It is also used on a volume basis for the concentration of a vapour in air. In this case the conversion to a weight per unit volume basis (e.g. mg/m^3) depends on the molecular weight of the chemical forming the vapour.

Abbreviations

Note that the abbreviation often derives from a chemical name that is no longer current, therefore some of the names below do not follow modern conventions.

AF-2	2-(2-furyl)-3-(5-nitrofuryl)acrylamide, a food preservative
AIDS	Acquired immune deficiency syndrome
ATA	Alimentary toxic aleukia
BHC	Benzene hexachloride, more properly hexachlorobenzene
CT	Computerised tomography
Ct	The product of concentration of a gas or vapour multiplied by the time of exposure – usually expressed as mg/m^3/min
2,4-D	2,4-dichlorophenoxyacetic acid or its esters
DDT	Dichlorodiphenyltrichloroethane
DFP	Diisopropylfluorophosphonate
DMSO	Dimethylsulphoxide
EPA	Environmental Protection Agency (in USA)
FDA	Food and Drug Administration (in USA)
GA	German code name for Tabun, or ethyl *N*,*N*-dimethylphosphoramidocyanidate
GB	German code for Sarin, or isopropylmethylphosphonofluoridate
GD	German code name for Soman, or pinacolylmethylphosphonofluoridate
GC-MS	Gas chromatography-mass spectrometry
GIGO	Guff in, guff out, or garbage in, garbage out, or any other modification you fancy
LC_{50}	A concentration of a chemical which kills 50% of a test population of organisms
LCL_0	Lowest concentration which has been reported to kill a test organism

LCt_{50}	The Ct (see above) which kills 50% of a test population
LD_{50}	The dose of a chemical which kills 50% of a test population
LPG	Liquefied petroleum gas
LSD	Lysergic acid diethylamide
MIC	Methylisocyanate
PAH	Polycyclic aromatic hydrocarbon
2-PAM	Pyridine-2-aldoxime methiodide
PBB	Polybrominated biphenyl
PCB	Polychlorinated biphenyl
PVC	Polyvinylchloride
SCUBA	Self-contained underwater breathing apparatus
SD	Standard deviation
2,4,5-T	2,4,5-trichlorophenoxyacetic acid or its esters
T-2 toxin	A toxin from moulds (mycotoxin) belonging to the chemical class of trichothecenes
TCDD	2,3,7,8-tetrachlorodibenzo-p-dioxin
TCP	2,4,5-trichlorophenol
TDI	Toluene diisocyanate
TLV	Threshold limit value
VDU	Visual display unit

1

Introduction: a problem defined

The growth of modern technology has given birth to at least three 'Ages' with profound social implications. The Atomic or Nuclear Age was created in the 1930s and early 1940s, to become dramatically obvious in 1945 at Hiroshima and Nagasaki. The benefits of the Nuclear Age are still being debated, the principal question being whether it can produce cheap power without endangering people and the environment. It also has brought the possibility of the total destruction of life on earth. The Electronic Age is just starting. This was helped along by the development of semiconductors in the 1940s, and has been boosted by the microchip technology which has made cheap but complex computers a commonplace. Since about 1980 we have begun to feel the social impact of the Electronic Age, but I believe that much more change is to come; we are just at the beginning of the social change due to this Age. Finally, we have what I have called the New Chemical Age, originating in the 1930s, but becoming of consequence since 1945. It is 'New' to distinguish it from the previous chemistry concerned with metallurgy and organic dye pigments that developed during the nineteenth century. This Age is defined more precisely in the next chapter. Although the New Chemical Age has been less dramatic in impact than the other two, its consequences have permeated society much more effectively than those of the other two, although the Electronic Age is now initiating profound changes in social organisation. What I mean is that the products of the New Chemical Age have become part of our everyday life, without us taking much notice of their appearance. Now we accept them totally. Note that I have excluded a possible Space Age; the only social effect of space technology has been as a stimulus to the Electronic Age.

If we have accepted the products of the New Chemical Age, why is there any need to comment on the social impact of chemicals? I believe there is a paradox here, in that although we have accepted and become totally reliant on the chemical products of the age, we are still very uncertain about many aspects of the manufacture, use and disposal of these products. We have quietly accepted major benefits, but are unable to assess clearly the associated technical problems. We are concerned about chemicals in our food, in the environment and in the workplace, about synthetic drugs, the factories which make chemicals, and the places in which they are stored. The situation is that we have adopted the chemical age without knowing its consequences. It may affect the future of the human species very profoundly, and the term 'chemical shock' is becoming fashionable to describe the situation. An alternative view is that the problems are relatively minor, capable of being controlled by judicious action now. We just do not have the data necessary to assess the problem; more importantly, the public does not have the background knowledge which would enable it to assess those data which are available.

I can give a few concrete examples of incidents that have aroused fear of chemicals. Recently, in 1984, the escape of vapour from a chemical plant in Bhopal, India, has claimed somewhere around 2000 deaths and many more injuries. In 1978 the tanker *Amoco Cadiz* was wrecked off Brittany releasing over 200 000 tonnes of crude oil onto the sea coast. Crude oil is the main feedstock for the synthetic-chemicals industry. Mothers who took the drug thalidomide during pregnancy produced children with grossly underdeveloped limbs. The chronic over-users of common pain-killing drugs occupy one fifth of the kidney dialysis machines in Australia, at a cost of $30 000/year for each patient. There was an epidemic of skin and nervous-system disorders in Japan in 1968, affecting perhaps 1000 people. This disease was caused by the consumption of rice oil contaminated by the chemical mixture known as polychlorinated biphenyls (PCB).

The above examples are clearly defined incidents in which the cause has been traced with a high degree of certainty to chemicals. There are a large number of less well defined problems which have also been ascribed to chemicals. The problems may range from imaginary to real, and the connection with chemicals may be fairly certain or very tenuous. Thus a greater proportion of people are dying from cancer now than 50 or 80 years ago. Is this due to greater contact with cancer-producing chemicals, or is it a consequence of a lesser proportion of deaths being due to infectious disease, i.e. a reflection of the medical (and chemical) control of infection? Is there an increase in the number of children born with deformities in certain rural areas, and is this increase (if it has occurred) due to herbicides? Our food contains chemicals as preservatives; are these chemicals harmful to us? There are, therefore, examples of problems and disasters which have a clear chemical cause, but there are also many in which a chemical involvement cannot clearly be seen, and may in fact not be there. You, as a responsible member of the public, are going to have difficulty in deciding such issues.

Let us develop another historical theme. In 1945 large numbers of servicemen returned to civilian life determined to build a better social order. Many eminent scientists were released from military projects, and the results of these projects could be put to constructive use. Radar, antibiotics, rocket technology, insecticides, semiconductors and plastics had potential civilian uses; the feeling was that science would now assist in the creation of a new, socially just and technically advanced society. Science was to belong to the people, and was to give them a better material life. During the following 40 years of uneasy 'peace' this expectation has faded. The young men 'demobbed' in 1945 and 1946 have seen the clarity of their social dreams become clouded and the aims of society confused. We now question the benefits that science was to bring and, to a large degree, there is less effort directed at bringing science to the people, to give them the ability to judge and appraise science for themselves. You are now saying that my statements must be nonsense for, surely, the new communications media are popularising science more than ever before. Television is a medium superior to any available before 1945, and science shows are popular television viewing. Popular, yes, popularisation certainly, but understanding, no. Television science is based on the 'Ooh, aah, fancy that' principle, and the techniques employed are largely founded on that of the commercial. If the BBC becomes privatised, and the last community-service station goes broke, then the science producer will have only one aim, to get the maximum number of viewers. Science popularisation depends on sensation and controversy, in newspapers, on radio or on television. There is no attempt to give basic understanding, no attempt to give you the tools to form your own opinion. You are getting an opinion thrust down your throat with all the glamorous trimmings of a cigarette commercial. Yes, I know I am generalising wildly, I know there are many genuine attempts to educate. Yet I think the main effort is to bring you to a conclusion, not to prepare you to form your own.

Thus many of the controversies that surround chemicals and their social impact and influence are presented to the public by interest groups that want to mould public opinion. The groups may be environmentalists, chemical manufacturers, food faddists, drug companies, religious organisations, government departments; all with a particular message. If you have not got the basic understanding of the technical questions involved, you will become the dupe of the best salesperson, the most attractive presentation. It is the aim of this book to give you some of the tools with which to fashion your own opinions.

Because during the last 40 years science has been judged by many to have failed society, I believe there has been a move towards obscurantism. People feel that a rational approach to their problems has failed, therefore they are drawn towards the irrational. This move away from reason is prominent in alternative medicine and dietary fads, which will be discussed later. I have tried, myself, to be rational in this book.

The title of this book claims an approach to social questions. However, I am not a sociologist, and emphasise that this is the approach of a biochemist towards a discussion of the social significance of the New Chemical Age. Ideally, a sociologist should also be approaching from the other side, so that we meet and fuse in a balanced judgement on technology and society. If anyone wishes to, go ahead with the complementary book.

A little definition is required now. 'Chemical' is, of course, an adjective that by usage has been transmuted into a noun. All tangible matter has both a physical and a chemical aspect so that, to the purist, the term 'chemical' is useless as it embraces all matter. Common usage limits and distinguishes 'chemical' from matter termed 'natural' (a suspect distinction, as we will see). There is also a distinction between 'chemical' and 'biological' (again suspect, as a living organism – being tangible – is then chemical). I will use the popular mode, so that by 'chemical' I generally mean something that has been made in a factory from less complex ingredients. Remember, however, that exactly the same chemical could very likely be obtained from a natural source. Alcohol can be made from crude oil and thus be a synthetic chemical, or it can be produced from malted barley by fermentation and thus be entirely natural. Penicillins can be made synthetically; it is economically advantageous to produce some by fermentation and then, if necessary, to modify chemically the initial products. Therefore we have always lived in a chemical world, what has changed is the type of chemical.

I have not attempted to discuss the topic of the impact of synthetic chemicals on the Third World, simply because this would be a major study in itself. The society being considered is that of the developed countries: Europe, North America, USSR, Australasia, etc. Some reference is made to other, developing, countries for particular purposes, but essentially it is the technically developed society that is being commented on.

The problem is therefore that people do not understand the chemical world, and the requirement is for objective information in language which is understandable. This book endeavours to fill that requirement. I do not wish to force any particular point of view on you, but to give you the tools to formulate your own. It is fairly obvious where I have expressed opinion rather than fact; think about the issues for yourself and read widely. The chapters are to an extent self-contained, but it is desirable to read them in order, particularly Chapters 4 to 7 which give the basic instructions.

2

The dawn of the Chemical Age

We are living in a Chemical Age, which is new, and brings with it novel problems. Because of the newness of the Chemical Age, people have no prior experience of these problems and therefore cannot assess them fully as to their severity and likely impact on society. This thesis is the starting point of this book and, in this chapter, I attempt to define the Chemical Age and to demonstrate that it is indeed new. We can use two bases of comparison to contrast modern society with previous ages: so-called primitive societies still existing today, or historical accounts of our own development.

If we are to believe Marshall Sahlins (1), the earliest human societies of hunter-gatherers deliberately avoided possessions, as these limited their mobility. They were able to enjoy the original 'affluent society' because their requirements were restricted and easily supplied. The Australian Aborigines fit into this cultural type. They are 'primitive' according to our concepts of society, but enjoy, or enjoyed, an intricate social and spiritual culture which was independent of the burden of possession and ownership which so dominates our traditions. Indeed, the one chemical that was an article of trade and also the most prized possession was red ochre (2) which was used in social and religious ceremonies for body and rock painting. This fits well with Sahlins' description of the hunter-gatherers' society; a great deal of leisure was available which was spent in social activities, rituals and so forth.

Other primitive societies had similar limited use for chemicals, apart from the great importance of salt, as exemplified by the extensive trade (the Saharan salt caravans) and its social significance. Primitive societies are now entirely penetrated by the chemical industry, and cannot give us much real information. For example, the remote African village may seem primitive, but the footwear is likely to be made from old car tyres, and the plastic container is ubiquitous.

When people settled down in neolithic times to dwell in one place, they were able to acquire possessions and to retain them. Mobility was not a necessity. Whether the people became the owners, or whether the goods and chattels enslaved them is a matter for cogitation. The result was that material things became much more prominent in society and have continued to grow in importance, particularly if you make a literal interpretation of television advertising. Chemicals began slowly to play a role in the development of this material culture. I think this can best be illustrated by considering the situation of medieval people in western Europe, then briefly comparing this with early industrial society, and finally contrasting both with the present.

What chemicals were used in medieval society? The answer is, quite simply, none at all if we are considering the majority of villeins, serfs and labourers that constituted this society. The excellent book by Rowland Parker (3) on the village of Foxton gives the dismal, unromantic, but no doubt true picture of the lives of the villeins and bordars who just survived in the medieval society. They lived in wattle-and-daub huts, furnished with wooden benches and boards, and ate from wooden platters and bowls. Food was cooked in crude earthenware pots. They wore the clothes they could make; homespun wool for tunic and leggings, leather for footwear. In brief, these people possessed what could be obtained from natural sources within the confines of their manor, and which could be fashioned by their own efforts. The Quennells' (4) picture of medieval life is less stark, but again emphasises the self-sufficiency of the manor as the economic unit. The latter reference lists the following as items bought outside the manor: tar, fish, furs, salt, iron, spices, silks and fine cloths. As far as the majority of the inhabitants of the manor were concerned, salt would be the only item they really had to have. This was necessary for salting pork, mutton or beef for storage over winter.

The requirements of the village were thus satisfied by local production. Many complex chemical processes were in fact carried out in trades such as textile dyeing or leather tanning, but the skills acquired by trial-and-error are those of the craftsman rather than of the technician. Thus if it were desired to dye a wool garment, then the dyes were obtained from local vegetable sources; yellow-brown from onion, purple from elderberry, etc. Such sources of dyes continued to be used until recent times. My mother bought some handwoven tweed in the West of Ireland in 1949 that was dyed in the two colours from the fuchsia flowers that grew plentifully in the hedgerows. Then, that was a survival of an ancient practice. Today, such a usage would be deliberately cultivated as a piece of preserved history. As another example, leather was tanned in the extract of oak bark, now replaced by chrome tanning. Bark for the tanning was therefore an important commodity and when it was found that wattle bark was a good source of tannins, the wattle trees of Australia and South Africa were exploited briefly, before chrome took over. I doubt if the chemistry of bark tanning has yet been completely described. Its use did not depend on theoretical knowledge but upon accumulated observation and practical trial. Some idea of the crafts that reflected this local self-sufficiency can be gathered from books like that of Manners (5), and the gradual decline of this rural life into modern (and materially better) times is recounted in oral history (6) or personal recollections (7).

The medieval manor ate what it could grow, and therefore was very susceptible to crop failure. There was little chance of any relief coming from outside, so that if your crops were devastated by insect or fungal pests, then you simply starved to death or became an itinerant beggar. The account of the potato famine in Ireland (8) describes the enormous effect on a nation of a fungal pest, in a society which was still, in many respects, medieval.

If we leave the medieval peasants firmly rooted in the soil, their vision extending the breadth of the manor, and their needs satisfied solely from its produce, does the following Industrial Revolution bring in a chemical age? Let us consider the industrial worker sometime around 1850 in a 'one-up, one-down' terraced hovel close to the factory or pit, in the soot and clamour of the city. The energy is mechanical energy from coal and steam, the materials are iron and steel. Chemistry as a science has been born, and the metallurgical knowledge of 3000 years has been greatly expanded. Yet, in my mind, this is not the Chemical Age. The stimulus for the New Iron Age that accompanied the Industrial Revolution was the development of new techniques to smelt iron in larger quantities, using coal and coke rather than charcoal. The construction of the Coalbrookdale Bridge (the first bridge of cast iron) in 1779 was a technological feat, not a triumph of chemical science. The extraction and refining of ores is a processing and modification of the same material taken from the ground; it is not a synthesis of something new. Thus in 1850 we had not reached the Chemical Age. The differences from medieval times are great: the source of energy is steam rather than man, animal, wind or water; the common material is iron from a large works; and the fuel is coal. These are not chemical changes in our sense.

I mentioned above that the science of chemistry had been born by 1850. Wöhler in 1828 had made urea, an organic compound of obvious biological origin, from silver cyanate and ammonia, two compounds recognised as being inorganic. This synthesis demonstrated that there was no absolute division between inorganic (including metallic) compounds and organic ones (principally compounds of carbon with hydrogen, oxygen and nitrogen). Organic chemistry became of industrial significance when W. H. Perkins made the first aniline dye (mauve, or aniline purple) in 1856 and founded the synthetic-dye industry. Meanwhile, the systematic basis of chemistry was being defined. The relationship between the chemical elements was established by Mendeleev in 1869. However, the rules for the combination of elements into compounds were not clearly defined until the early twentieth century, by G. N. Lewis and others.

If we now move from the 1850s to the 1980s, the scene is entirely chemical. Modern people are dressed in clothes of nylon and polyester, or a mixture of those with natural fibres. Their houses are floored with carpets of a wool and nylon mix, the kitchen with vinyl sheet. The kitchen benches are covered with synthetic laminate (Formica, Laminex, etc.); if the furniture is wood, it is coated with a polyurethane varnish. If it is not wood, it is made of perspex or polypropylene. The car structure and engine remains metal, but the whole interior is vinyl. Food is supplied clean and free from the ravages of insects and rodents, because these enemies are controlled by synthetic pesticides during growth and storage. The supermarket shelves are stocked with an arsenal of chemicals, principally now in aerosol cans. There are insecticides for the home and garden, cleaning preparations of all types, sprays to make starching and ironing easier, sprays to recondition teflon-lined frying pans, sprays for every conceivable purpose. Then there are the bulk chemicals, exemplified by the laundry powders and liquids. This is truly the Chemical Age.

What is important to recognise is that this Chemical Age is very new, and therefore presents very new problems to our society. The War of 1939–45 probably separates the Chemical Age from the New Iron and Steel Age. The synthetic chemicals can be divided into four groups: the structural chemicals, the pesticides, the synthetic drugs and the process chemicals. Examples of this classification are given in Table 2.1. The structural chemicals (in common terms the plastics and synthetic fibres) were developing slowly before 1939, but received a great boost during the war. Perspex, for example, became important for aircraft canopies, and nylon for parachutes. Volume production began with the post-war economic recovery. Similar considerations apply to the pesticides. Organophosphate insecticides were being discovered just prior to 1939. The war nearly caused them to be diverted to military use (see Chapter 10) and certainly increased interest in their chemistry. The phenoxyacetic acid herbicides (2,4-D; 2,4,5-T) had a partly military origin, and DDT had its first major use among troops in controlling typhus in Naples in 1943–4. The development of process chemicals was apparent from the advertising wars fought by detergent manufacturers. Immediately after the war the pattern of development was set. The chemists were able to diversify the number of useful chemicals very rapidly and the chemical engineers had no problem in expanding production. The Chemical Age is therefore at most 40 years old.

It should now be clear what I mean by the Chemical Age. It is the period in which synthetic chemicals have occupied a near-dominant place in our economy and daily life. Synthetic chemicals are those which have been formed by industrial processes from precursors entirely different in properties from the final product, as distinct from chemicals and substances obtained by an extractive process and refined. The synthetic chemicals are produced in large installations (Fig. 2.1) which themselves must have an influence on society and the environment in which we live.

The development of the synthetic drug industry has had a slightly

Table 2.1. *A functional classification of synthetic organic chemicals*

Class	Description	Examples
Structural chemicals		
Bulk plastics	Used for implements, furniture, plumbing, etc.	Perspex, polystyrene, PVC, nylon, melamine, polythene
Synthetic fibres	For use in textiles, carpets, ropes, etc.	Rayon, nylon, terylene
Pesticides	For control of pests, such as insects, weeds and molluscs (snails)	Parathion, carbamates, coumarins, 2,4,5-T, metaldehyde
Drugs	Chemicals for the treatment of disease, or which have a marked effect on the body.	Antibiotics, anti-depressants, anti-hypertensives, vitamins etc.
Process chemicals		
Industrial	To facilitate a process or to modify it, without forming a significant proportion of the final product	Wetting agents, mould-release agents, sterilants, additives for many purposes
Domestic	For a variety of uses, often supplanting substances from natural sources	Detergents, polishes, disinfectants

different source from the mainstream of chemistry, having origins also in herbalism and natural extracts. This is described in Chapter 9.

Before we attempt a detailed, quantitative look at the chemical world, I will briefly sketch the chemistry of synthetics, for the general reader. Let us make nylon. This chemistry hinges on the properties of the carbon atom. It has four bonds available to attach to other atoms (*a* of Fig. 2.2). If we attach four hydrogen atoms, we get the gas methane, as in (*b*) which is the firedamp of the miner and a component of natural gas. We can also attach the carbon atom to other carbons to form a chain (*c* of Fig. 2.2), 16 carbon

Fig. 2.1 A modern chemical plant. The Olefin 6 Naphtha Cracker of Imperial Chemical Industries' P & P Division, Wilton, Teeside (photograph courtesy of ICI Plc).

Fig. 2.2 The synthesis of a polymeric fibre, Nylon 66; see text for details.

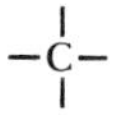

(*a*) Carbon atom

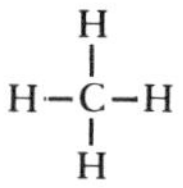

(*b*) Methane

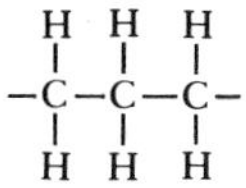

(*c*) Carbon chain

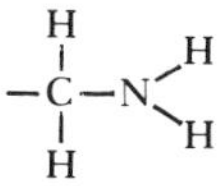

(*d*) Organic amine

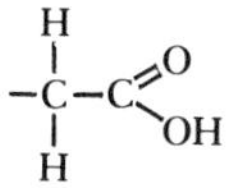

(*e*) Organic acid

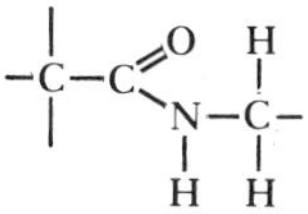

(*f*) Amide link

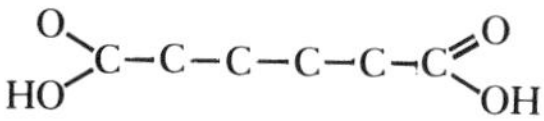

(*g*) Adipic acid

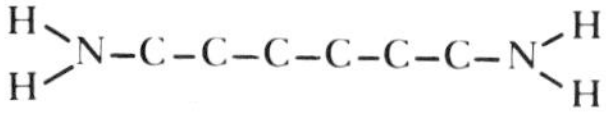

(*h*) Hexamethylene diamine

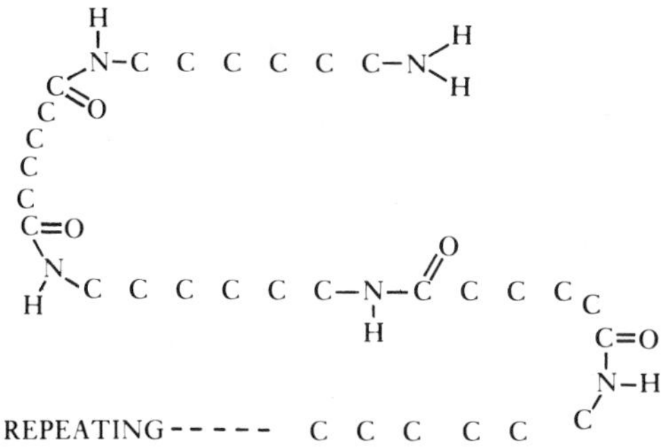

REPEATING - - - - -

(*i*) Nylon 66

chains are common in oils and much longer ones are possible. Other elements can join to the carbon atom, as shown for nitrogen in (*d*) to form an amine, or for oxygen in (*e*) to form an acid. It is not easy to form a direct bond between carbon and carbon. It is easier in industry to start with a long carbon chain (from crude oil) and break it down to the required carbon chain length. However, we can join carbons up through other atoms. Thus an organic acid and an amine will link up to form an amide bond, as in (*f*). So we can join small molecules up to form longer ones. Why not join longer acids and amines together to form even bigger components? Come to that, why not have a joining group at each end of the molecule, so that the small molecules can be joined in long, repeating chains? All right, we will take adipic acid, (*g*), and hexamethylene diamine, (*h*), and join them up. These two will join up repeatedly to form very long chains (of hundreds of repeating units). We have just made Nylon 66. Unfortunately the chemists of Du Pont took out the patents in the 1930s, so we cannot profit from this exercise, except intellectually.

Nylon is made up of smaller units in repeating chains. The units are known as monomers, and the big chain as a polymer. The chemistry of structural plastics and fibres is therefore known as polymer chemistry. Pesticides are not polymers; the molecules are relatively small, of molecular weights (of a few hundred) as opposed to hundreds of thousands for polymers. Molecular weight refers to the mass of the compound to that of the hydrogen atom as unity. Carbon is 12, so that methane (Fig. 2.2) has a molecular weight of 16, adipic acid 146 (oxygen is 16).

Let us now make a pesticide, 2,4,5-T. We start from benzene (*a* of Fig. 2.3) which can be obtained from crude oil. Six carbon atoms are joined in a circle which utilises three of the bonds of each carbon atom, so that only one hydrogen can attach to each carbon. A simplified rendering of benzene is shown in *b*; the presence of hydrogen atoms is implied. We can substitute an oxygen and hydrogen group for one single hydrogen, which gives us phenol (carbolic acid, the old disinfectant). By various means we can also substitute three of the hydrogens with chlorines, in set positions, to give 2,4,5-trichlorophenol or TCP (*d*). The acetic acid shown in (*e*) is the main constituent of vinegar. The phenol derivative and the acid can be combined to form the 2,4,5-T, shown in (*f*). This can be further modified to vary its solubility and persistence as a herbicide. So we have now made 2,4,5-T. If that were all we had made, we would be quite happy. Unfortunately, it is possible for two molecules of trichlorophenol to unite to form 2,3,7,8-tetrachlorodibenzo-p-dioxin; TCDD or commonly 'dioxin'. This very toxic contaminant of 2,4,5-T has been the cause of many social problems and much debate (see Chapters 11 and 13).

It is wrong to assume that chemicals as supplied are pure. Dioxin is the most dramatic illustration of a problem contaminant, but we must always be aware of such problems in relation to all chemicals. Polymers (plastics) may contain monomers that have not united into chains, a problem example being vinyl monomer which has some toxicity and may occur in vinyl polymers. Furthermore, various chemicals are added to monomer mixes to initiate the joining up of the monomers. Traces of such unconsumed chemicals may be a problem. The ideal polymer is all polymer, unfortunately this ideal state is never attained. Other chemicals may be added deliberately as plasticisers to modify the properties of the polymer. Products made of polyvinylchloride (PVC) usually contain a high proportion of phthalate esters to make the plastic flexible. These esters appear to have some toxicity; there was a worry that small amounts of these compounds might be getting into blood from the PVC bags and tubing

Fig. 2.3 The synthesis of a pesticide, the herbicide 2,4,5-T; see text for details.

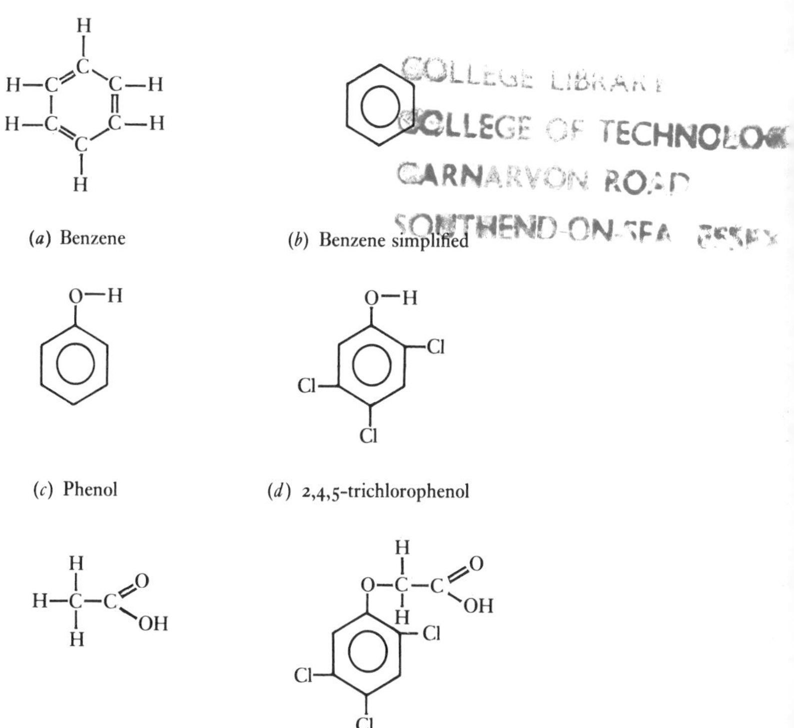

(*a*) Benzene (*b*) Benzene simplified

(*c*) Phenol (*d*) 2,4,5-trichlorophenol

(*e*) Acetic acid (*f*) 2,4,5-trichlorophenoxyacetic acid (2,4,5-T)

used in transfusion procedures. There is still some uncertainty as to whether such contamination is an actual danger to the patient, but it is obviously good policy to choose polymeric materials which are as free from small molecules as possible. Large molecules are less of a problem because they are not soluble in water or other solvents and do not diffuse through other materials. They stay where you put them, which is reassuring. Small molecules can turn up in unexpected places.

Finally, in this chapter, we will try and get some quantitative idea of the increase in production of synthetic chemicals. This is not easy, and figures by themselves are not too informative. You can get a reasonable idea of the extent and variety of production by scanning through copies of *Chemical & Engineering News* (American Chemical Society) or *Chemistry in Industry* (UK Society of Chemistry in Industry). Figure 2.4 taken from a paper by Davis & Magee (9) shows the rapidly increasing rate of production of synthetic organic chemicals. Production levels have doubled every 7–8 years since the late 1930s, to reach 350 billion pounds per year by 1980. Lately it has oscillated following a sharp recession in 1981 (10). The latter

Fig. 2.4 The growth of production of synthetic organic chemicals and of total chemicals and allied products, annual value added. (Reprinted from (9) by permission of *Science*; copyright 1979 by the AAAS.)

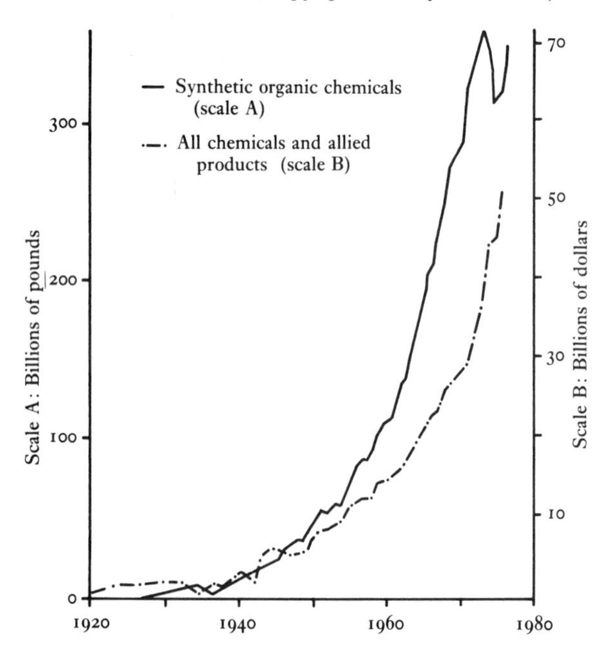

reference gives detailed figures for the US chemical industry. In 1981 the US production of thermosetting resins (epoxy, polyester, melamine, etc.) totalled 5 billion pounds; that of thermoplastic resins (polyethylene, polypropylene, nylon, PVC etc.) was 28 billion pounds (US billion, 10^9). Synthetic fibres (including rayon, nylon, polyester, etc.) totalled 9.8 billion pounds. These figures are rather indigestible, but what is important is the growth in production, as shown in Fig. 2.4.

It is not only growth in quantity that has occurred, but also in the variety of chemicals developed for specific purposes. Table 2.2 lists 13 classes of additives which are used in conjunction with plastics, that is the first item in Table 2.1. Some of the names for the additives are immediately comprehensible, others are intriguingly obscure. 'Additive master batches' are premixes of additive and base chemical, sold to the plastics manufacturer for further dilution into the base chemical during production. Some additives form a sizeable part of the finished product, and are a structural part of it. Plasticisers are examples; phthalate plasticisers may form 30% or more of the PVC product. Other additives are put in to accelerate or direct the formation of the polymer from its constituent monomers, e.g. the urethane catalysts. These may be consumed in the production process and not appear in the product. Many of the compounds include metals in their formulation, so that added to the possible toxicity of

Table 2.2. *The variety and consumption of additives for plastics in the United States*

	US consumption (000 tonnes)		
Type	In 1983	In 1984	% increase
Plasticisers	657	718	9
Flame retardants	190	206	8
Inorganic pigments	126	135	7
Synthetic dyes	41	43	5
Heat stabilisers	37	39	5
Blowing agents	5.3	5.6	6
Urethane catalysts	2.4	2.6	8
Antimicrobials	No figures (low consumption)		
Antistats			
UV stabilisers			
Impact modifiers			
Antioxidants			
Lubricants			
Additive masterbatches			

Adapted from (11).

the synthetic chemicals is that of these metals, which include tin, antimony, cadmium and lead.

Each of the 13 classes in Table 2.2 has many diverse chemicals in it. It is difficult to know how diverse, as most are known by trade names. A lot of ferreting around is necessary to find out whether two products from different companies are chemically distinct, or whether they include the same active principle in varying formulations. Therefore a chemist finds it difficult to wade through the sea of trade names and jargon to establish exactly what is being used in industry. The chemist must specialise in one small area of the technology; the layman has no hope. The publication from which some of Table 2.2 was drawn (11) states that over 200 new additives were brought to market in 1984. The great majority of these 200 would be formulations of existing compounds. Added to this will be new fibres, pesticides, drugs and process chemicals (many of the items in Table 2.2 belong to this latter class). The variety to superficial view is enormous, and even when reduced to eliminate varied formulations of the same chemical, the variety is sufficient to make control very complex.

Table 2.2 also shows the expected increase in US consumption from 1983 to 1984, which I added out of general interest. An average increase of 7% presumably indicates the rising production of the bulk plastics to which all these additives impart some desirable property.

Whether we like it or not, we are in the chemical world. What I want to do in this book is to explain the benefits and problems that chemicals bring, and thus to help explain their impact on society. It must be stressed that the chemicals of the Chemical Age are the synthetic, organic chemicals, compounds of carbon. Metallurgy has a chemistry of its own, developed over the last 5000 years, and tremendously increased in complexity in recent years. Aeronautical science and space exploration has been made possible by the development of metallic alloys and ceramic materials. I do not wish to decry the achievements of the metallurgists or the inorganic chemists, it is simply that I am using the term Chemical Age in a restricted sense to apply to synthetic organic chemicals. My reason is that this latter group has given rise to the most discussion in public, and so the most doubts about their effects on society. In Fig. 2.5 are shown the traditional roots of chemistry, in animal and vegetable products, and metals. Metallurgy develops separately, but all the other roots are drawn into some association with synthetic chemistry, or are entirely superseded by it. Thus modern drug therapy owes a lot to herbal medicine (Chapter 9), but manufacture of drugs today is largely from synthetic chemicals. Similarly, dyestuffs are synthetic; their natural precursors from vegetable sources are used only in textile craft.

As a moderately detailed account of those chemicals which most impinge on our society, I can recommend the book by Selinger (12), without entirely subscribing to all his comments.

We have thus, for the last 40 years, dwelt in the Chemical Age, a novel era of material benefits and unknown dangers.

References

1. Sahlins, M. (1974). *Stone Age Economics.* London: Tavistock Publications.
2. Mountford, C. (1967). *Brown Men and Red Sand.* Sun Books edn, p. 74. Melbourne: Sun Books.
3. Parker, R. (1976). *The Common Stream.* St. Albans: Paladin.
4. Quennell, M. & Quennell, C. H. B. (1945). *A History of Everyday Things in England.* Part 1: *1066 to 1499.* 3rd edn, p. 113. London: Batsford.
5. Manners, J. E. (1974). *Country Crafts Today.* Newton Abbot: David & Charles.
6. Evans, G. E. (1975). *The Days That We Have Seen.* London: Faber and Faber.
7. Page, R. (1975). *The Decline of an English Village.* London: Davis-Poynter.
8. Woodham-Smith, C. (1962). *The Great Hunger.* London: Hamish Hamilton.
9. Davis, D. L. & Magee, G. H. (1979). Cancer and Industrial Chemical Production. *Science*, **206**, 1356–8.
10. Anonymous (1982). Production by the U.S. Chemical Industry. *Chemical & Engineering News*, June 14, 1982, 32–8.
11. Anonymous (1984). Special Report: Additives. *Modern Plastics International*, September 1984, 31–52.
12. Selinger, B. (1975). *Chemistry in the Market Place.* Canberra: Australian National University Press.

Fig. 2.5 The roots of synthetic chemistry in metallurgy, herbal medicine and natural product chemistry.

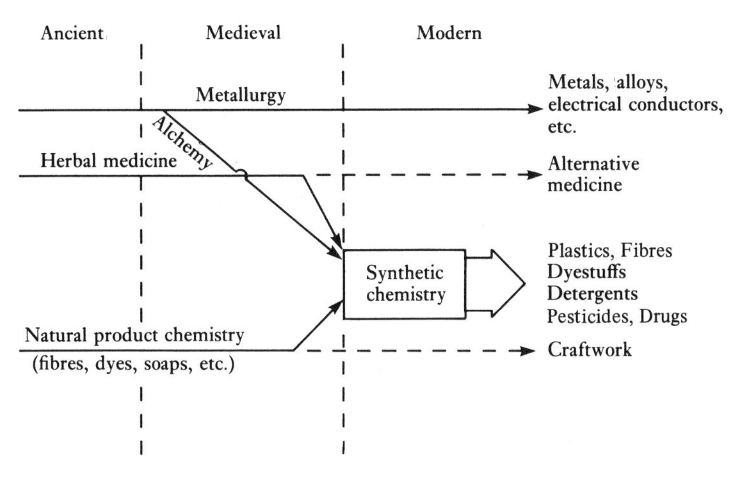

3

The benefits of the Chemical Age

The Chemical Age has not been forced upon a reluctant population by the machinations of technocrats or by an industrial conspiracy. It has been enthusiastically supported by the common people in developed countries who willingly buy the many products available. This can be seen at the domestic level by a visit to a supermarket, as mentioned in the previous chapter. Insecticide cans are bought for the extermination of any insect found in the home, whether harmful or not. Cleaning products keep the home bright and sparkling, pesticides and fertilisers produce healthy growth in the garden. The chemical manufacturer cannot be blamed for whatever excessive usage of chemicals may occur; he may stimulate demand by advertising, but the customer finally makes the decision to buy. The reasons why domestic products are so widely purchased are complex. One can argue that the desire to have the home spotless, the toilet completely sterile and the garden free of bugs stems from a neurotic preoccupation with the home caused by a society that idealises the housewife as the domestic guardian. Perhaps this is so. Other practices can be questioned. Gardeners in my neighbourhood have to buy fertilisers because they burn their leaves in the gutters and the ash goes down the storm-water drains. Composting of the leaves would conserve the existing garden nutrients. Whatever the deeper reasons for the demand, if the aim is to attain a clean home or neat garden, then chemicals offer a benefit. More evident benefits are seen from the chemicals the consumer buys as finished products and those which are used in industry. The four categories of chemicals given in Table 2.1 (that is, structural chemicals, pesticides, drugs and process chemicals) are considered in turn.

The bulk plastics are essentially in competition with metals, wood and other materials of natural origin. A detailed consideration of properties of the contending materials, in the particular application that is desired, must be made. Plastics will not necessarily oust all other materials. Consider PVC, for example. It is ideal for drainage systems in domestic plumbing as it is cheaper than lead, copper or brass and easier to work with. Also, it is superior to earthenware drains as sections of flexible PVC pipe can be very long and can be sealed. Earthenware sections have to be short, to allow for some movement, and the joints are often then penetrated by tree roots. As the insulating cover for electrical wiring PVC is also ideal, being much more durable than natural rubber. However, the use of PVC for gutters and downpipes is a doubtful proposition. It is certainly sold for that purpose, but its use above ground exposes it to that old enemy of plastics, ultraviolet radiation from sunlight. My gutters are aluminium, factory-coated with an enamel of unknown composition. The only plastic that will remain clear for long periods when exposed to sunlight is Perspex, which is expensive. Therefore glass will remain as a glazing material despite its brittleness. Similarly, if we compare materials for domestic crockery, the contenders are ceramics and melamine (other plastics are too easily scratched). Melamine is used for some circumstances where breakage may be a big problem, but it is also expensive, so that ceramic materials are still more commonly in use. Of course an element of tradition persists in the more expensive ceramic products; Spode, Royal Copenhagen and Mikasa can offer design and craftsmanship not yet developed for plastic products.

Aeronautical engineering has up to now been the province of the metallurgist, but composite organic materials (i.e. glass or carbon fibres embedded in a plastic matrix) are beginning to displace metals, and the aircraft of the future will be fabricated from composite sheets joined with adhesive, rather than from metal alloy sheets rivetted together as at present. The relative advantages and disadvantages of plastics versus metals are given in Table 3.1. It can easily be seen that metals are superior for one application, plastics for another. The new aircraft may have a plastic composite fuselage and wings, but the engine components will be of metal or ceramic, since plastic will not stand up to the high engine temperatures. The benefits of bulk plastics are thus undramatic, depending on a careful appraisal of cost and suitability. More uses will be found for them, and we will find them to be yet a greater part of our environment.

The bicycle tyre levers and tool box shown in Fig. 3.1 illustrate the result of good design applied to modern plastics. The new product is lighter than the metal items it displaces, and is not subject to corrosion. Other technologies make demands on the polymer chemist; circuit boards for the Electronic Age have to meet exacting requirements, which plastics can meet (Fig. 3.2).

If we look around our environment we too often see obvious signs of one usage for structural plastics, that is we see the disposable plastic container improperly disposed of. The modern packaging revolution is largely dependent on polythene, polystyrene and PVC. The benefits of this are in cost savings to the distributor, who saves on the labour costs of one-off packing at the point of sale, and in convenience to the purchaser, who can select a pack from the shelf and know that it contains a product of consistent quality well protected from the environment. The disadvantages are related to the increased cost (to the consumer) of the pack, and to the disposal of the empty container.

Synthetic fibres (second item of Table 2.1) have found their most satisfactory applications as mixtures with natural fibres. Such mixtures retain the comfort and handle of the natural fibre and receive extra strength and wear from the synthetic. If new polymers are invented with superior qualities, then natural fibre may be largely displaced. The benefits of the synthetics are economic: lower cost and longer life. Natural fibres confer the aesthetic benefits (compare with ceramics for tableware as mentioned above).

The pesticides have more dramatic benefits, since there was little in use for the purpose before the Chemical Age. The two benefits of pesticide use are increased food availability and better health. The increase in available food results from protection of the crop in the field, the removal of competing plants (by herbicides) and the protection of the harvested crop

Table 3.1. *A comparison of the advantages and defects of metals versus plastics*

	Metals	Plastics
Advantages	1. Heat resistant	1. Not readily degraded at normal temperatures
	2. Hard surface	2. Light in weight
	3. Insoluble in most liquids	3. Quiet, e.g. nylon for gears
	4. Non-flammable	4. Fabrication often easy, may be stuck together
Defects	1. Corroded by water and salts	1. Deteriorate in sunlight
	2. Noisy – as structural panels or moving parts	2. Rapidly degraded at high temperatures
	3. May be hard to fabricate	3. May be flammable
	4. Jointing requires high temperatures (welding) or mechanical fasteners (rivets, bolts, etc.)	4. Surface soft easily scratched
		5. May be dissolved by liquids, e.g. cleaning fluids

in the store. It is easy for the city dweller to ignore the benefits of crop protection; we have not seen a countryside completely stripped by locusts, our apples from the supermarket are not eaten out by the codling moth, our grain products are not full of weevils (Fig. 3.4) or rat droppings. Why have we not seen these things? Because the locusts are watched, early signs of a plague are recognised, and the young hoppers are attacked on the ground with chemicals. Because the codling moth is killed by pesticides in the orchard before it can lay eggs. Because our grain is stored in granaries in which insect infestation is deterred by chemicals, and in which invading rodents are also killed by chemicals. Humans have no natural right to the bounty of nature. We have to fight for it against all the other children of nature, the weevils, cockroaches, rats, mice, locusts, moulds, blights,

Fig. 3.1 This bicycle tyre lever set and tool box show the advantages of good design allied to modern materials. The tyre levers are injection moulded from 'Maranyl' A192, a glass-reinforced nylon from ICI, and the tool box from 'Propathene' polypropylene grade LYM 43 (photograph courtesy ICI Plc).

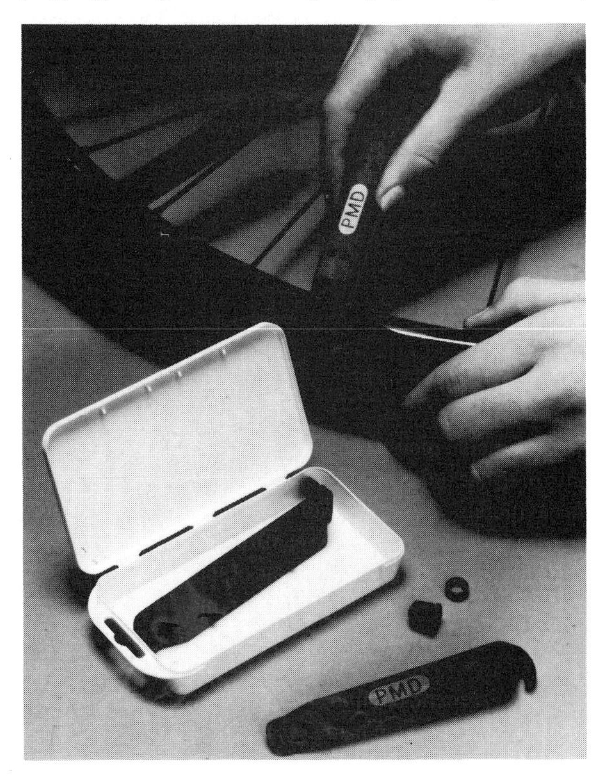

viruses, eelworms and whatever, that also wish to flourish in the world. It is war, the weapons are largely chemical. To fail is not merely the distaste at finding a grub in our apple, it is to die of starvation. We, the overfed, as one quarter of the world's population may find this hard to understand. The Bangladeshi peasant has a much more realistic view of life. In the future, if we are to feed and provide a more equitable style of life for the inhabitants of this world, then I cannot see how this can be done without chemical pesticides.

Pesticides are a benefit to health because many diseases are borne by insects or other vectors. It is better to kill the vector carrying the disease organism, rather than to wait until that organism has been transferred to a human and the sickness has commenced. Mosquitos are good examples of vectors. The malaria organism, as everyone knows, is carried by the anopheles mosquito. Yellow fever is borne by a related species. Various

Fig. 3.2 The electronic age demands new materials from the synthetic chemist. These circuit boards are moulded from 'Victrex' polyethersulphone (photograph courtesy of ICI Plc).

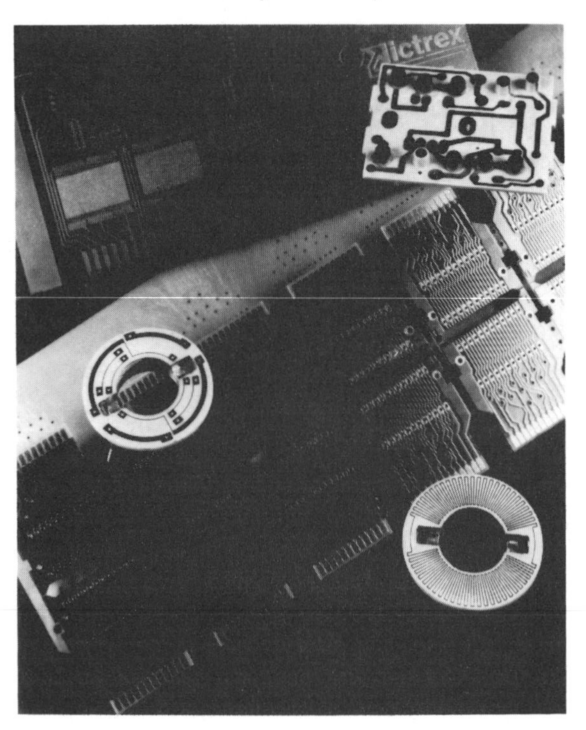

forms of encephalitis are carried by mosquitos, often in limited localities, but of importance because of the seriousness of the disease. Mosquito-borne encephalitis occurs in areas such as Chiangmai (Thailand), the Murray Valley (Australia), and epidemically in the USA as St Louis encephalitis (e.g. at Houston in 1964). In the late 1950s there was a real hope that malaria could be eradicated from the world by chemical control of the mosquito, and the World Health Organisation set up a program to achieve this. Unfortunately, for various reasons, this program has failed. The malaria organism is still with us and has developed resistance to some of the drugs used to kill it in infected humans. The disease that has been eradicated is smallpox, by means of a biological control measure (vaccination). Pesticides have failed to eradicate malaria, but still have an

Fig. 3.3 Synthetic fibres are a major proportion of total chemical products (photograph courtesy of ICI Plc).

important role to play in controlling it. Many other diseases can be controlled by the appropriate use of insecticides.

The benefits from the use of pesticides are given by Whitten (1) in a book written largely as a counter to Rachel Carson's *Silent Spring* (2). To my mind the benefits from the use of pesticides are not in doubt. What is debatable is whether or not we are making the best use of them, and whether we are creating long-term problems by incautious use of them at present. We have to balance the benefits against possible ill effects. Quite a proportion of the rest of this book considers the latter effects. Nor are chemicals the only means of control of pests. Because this book is about chemicals only, the reader must not get the impression that I do not see the value of other control methods. The malaria/smallpox comparison above is instructive.

The development of synthetic drugs has also brought us benefits, along with a number of social problems. Drugs have been developed (antibiotics) which will entirely destroy disease organisms in the body, with assistance from the body's natural defensive mechanisms. Others do not cure disease,

Fig. 3.4 Insects can rapidly spoil foodstuffs if not controlled. Here the flour beetle (genus *Tribolium*) has infested wheat grain (photograph courtesy of Ministry of Agriculture, Fisheries and Food, Crown Copyright).

but control it and thus improve the quality of life of the sufferer (e.g. antidepressants, antihypertensives). An improvement in human health has been brought about by a complex group of factors, of which drugs are one. These points are further discussed in Chapters 8 and 9.

The benefits of what I have described as process chemicals (Table 2.1) are much harder to define. The group includes a miscellaneous array of chemicals with many different uses: wetting agents are used in the textile and paper industries among others; emulsifiers are used in the formulation of pesticides; ethylene glycol is the antifreeze in your car radiator; many synthetic chemicals are used in hospitals as disinfectants or bacteriostatic agents in fluids for injection. Synthetic detergents have displaced soaps in the laundry and have posed new problems for the sewerage-treatment plants. Currently, one of the most controversial process chemicals is tetraethyl lead, which is added to petrol to increase the octane rating and is, thereby, very efficiently dispersed around every city (as its combustion products). The benefits of these chemicals are largely to be expressed as increased 'convenience', a term which is difficult to render down into concrete benefits whether financial or social. In some cases, such as for disinfectants, the benefits are more obvious. The decrease in childhood mortality and hospital-acquired infections, commencing around 1900 (see Chapter 8), predated the acquisition of antibiotics and was due partly to the introduction of disinfectants. The older ones based on phenol (carbolic acid) have been replaced with more elaborate synthetic chemicals which are safer and more effective. Some of the bacteriostatic agents (benzalkonium chlorides) are also widely used in backyard swimming pools, with no obvious ill effect. In summary, it requires a detailed bookkeeping exercise to determine the benefit of any one of these chemicals. Each could be dispensed with, without a great loss to society. The advantage of each is related to the greater cost of an alternative process; a judgement has to be made based on all economic and social circumstances.

It is difficult to find meaningful quantitative data to express the benefits of chemicals. For structural chemicals and process chemicals the figures are hidden in industry; obviously the manufacturer finds a benefit in financial terms but just how much is difficult for outsiders to know. For pesticides, the benefit in terms of increase in crop production is given variously as between 10% and 50%. Once again, this is difficult to determine accurately, as other improvements in agriculture have occurred concurrently with the introduction of pesticides (1). If you are still sceptical about the advantages of pesticide use, then try to find the answers to the following questions for yourself.

1) What proportion of the harvested cereal crop is destroyed by pests before consumption in India and Australia?
2) How many pests attack rice growing in paddies, and what reduction in output occurs if these are uncontrolled?
3) To what extent has the use of molluscicides in Egypt controlled the disease bilharzia (schistosomiasis)?
4) What was the cost to the US cotton-growers of the crossing of the Rio Grande by the cotton bollweevil in 1892? How effective is control by pesticides?

The answers to these questions will not show pesticides to be the universal solution to all problems but will, I think, illustrate a solid degree of benefit in their use – wisely.

It is even more difficult to make a quantitative estimate of the benefits derived from synthetic drugs. You could attempt to do it by using parameters such as the cost of an average stay in hospital, the value of extending life by x years, the reduction (if any) in sick time off from industry. However, you always come up against problems such as: 'What is the value in dollars of keeping a 54 year old woman alive? or: 'What is the value of relieving the irritation of a chronic skin sore by cortisone treatment?' Such questions do not have quantitative answers; they must be judged by social values.

In the end, the recognition of benefits from chemicals is a value judgement which you must make for yourself. You must examine carefully all aspects of life which are influenced by synthetic chemicals, then translate these strictly material aspects into wider values of society. All I wish to say is that you should do your best to assess all the ways in which the Chemical Age has influenced our society. I do not wish to enter the wider debate about whether such influences and the concomitant changes are good or bad, that is for you to decide. The title of this chapter betrays my own conclusion but I will not seek to defend it here. We may use the development of pesticides as an example of the distinction I am seeking to make. That pesticides have increased the total production of foodstuffs is, I believe, proven. This judgement falls within the sphere of discussion in this book. Whether or not this increased production of foodstuffs has benefited human society in its widest sense is a question I am not discussing. Thus whether pesticides have purely increased the stockpiles in the granaries of Australia and Canada, or really helped us to feed the Bangladeshi peasant, is a fascinating question but extremely complex and not in the province of the present book.

In this brief chapter I have sketched the advantages of the Chemical

Age. This does not mean that I see no disadvantages; in fact many will be discussed later. Next I have to describe some of the basic principles of chemical toxicity, a knowledge of which is essential to anyone wishing to follow the current debates on the safety of chemicals.

References

1. Whitten, J. C. (1966). *That We May Live*. Princeton: Van Nostrand.
2. Carson, R. (1963). *Silent Spring*. London: Hamish Hamilton.

4

Toxicity and dosage: the varying shades of grey

The concept of a poison is well understood, or at least people believe they understand it. A little reflection may cause them to think again. Crudely, a poison is something that kills or harms someone, like cyanide or deadly nightshade. There is more to it than this, however. How do you define a poison, if in some cases a substance (e.g. alcohol) may cause death, and in other circumstances may be relatively harmless? The adjective used scientifically to qualify a poison is 'toxic'; 'toxicity' is then the property which such a substance displays. Contrary to popular belief, the fact that a substance is toxic does not mean it will kill people. A toxic chemical is one which causes some observable detriment to a living entity: death is not necessarily implied. Many people become intoxicated with alcohol, but few die from it. A toxic chemical may cause death or a skin rash, blindness, deafness, disease of the liver or kidneys, or a multitude of other effects, serious or trivial. It is the purpose of this chapter to explore the concept of toxicity and seek to give it a definite meaning.

If someone doses himself with 1 g of arsenic trioxide, he will most likely die. If he doses himself over a number of years with an excess of good, wholesome food, he may die of heart failure or some other condition partly due to his excessive eating. Clearly then, both arsenic and good, wholesome food are poisons. If you do not agree, tell me what the fallacy is in the above argument. In fact there is no real fallacy, what is necessary is a qualification of the term poison. To say a substance is poisonous is not meaningful unless a dose is also stated. Thus 1 g of arsenic will kill a man, 1 mg will not. Adequate food will maintain a person in good health; too much food or a poorly balanced diet will cause conditions which may prove to be very unfavourable. The question 'Is this chemical poisonous or not?' is meaningless until a dosage level is stated. The more sensation-minded writers for the popular press tend not to worry about such qualifications. 'A fire today at a suburban chemical factory caused the release to the atmosphere of two kilos of a poisonous chemical'. This could be a cause of grave alarm if the chemical was dioxin, or of no consequence at all if it were methyl salicylate (oil of wintergreen). The toxicologist is not being deceitful or dissembling when failing to answer directly the question 'Is this a poison?'; it cannot be answered 'Yes' or 'No'.

I once saw a report in a newspaper about a biochemist who had found that the herbicide 2,4,5-T caused degenerative changes in brain cells. This adds further to the public concern about the safety of using this herbicide in agriculture. However, two vital facts are necessary before the report can be considered relevant to this public concern. One is the dosage of 2,4,5-T to which the brain cells were exposed and the other is whether this dosage level is likely to be achieved in the brain of humans when the herbicide is used to spray blackberries, or whatever other normal use is made of it. The first fact could be obtained from the researcher, but the journalist obviously found it irrelevant or meaningless to his readers. The second question may not have an answer as yet. A precise answer would be very difficult to obtain, but a range of likely concentrations should be calculable using some reasonable assumptions. Without this latter information, the researcher's observations are irrelevant to the agricultural use of 2,4,5-T, whilst having real value in the study of mechanisms of brain toxicology.

How can we express the amount of substance administered, which produces the toxic effect? Firstly, there must be a precise unit of quantity, such as mass for solids or volume for liquids. If the chemical is known to be impure, then some indication of the purity is necessary. This applies particularly to semi-purified natural products that may be obtained in different batches, each containing a different proportion of the active principle. Each batch will show a different toxicity, which needs correction to relate it to the toxicity of the chemical of interest. Obviously it is preferable to work with pure compounds only; this is not always possible, particularly so when a new type of chemical is being investigated.

The dose given needs to be related to the size of the recipient. Thus a tolerable dose for an 80 kg adult may be fatal to a 10 kg child. Body weight is the commonest expression of size, so that dosage is usually expressed as weight of chemical per unit body weight; milligrams per kilogram (mg/kg) being the units of most general usefulness. These equate to parts per million (ppm). Body surface area is sometimes thought to be a more useful measure of body size, particularly when extending toxicity data from small animals to humans. This leads to units such as grams per square metre (g/m^2). The body surface area for adult humans varies between 1.5 and 2 m^2; for rabbits it is about 0.15 m^2.

The quantity of a chemical required to produce an effect on the body is dependent on the way it is administered. Therefore, toxicity data need to be qualified by an indication of the route of administration. Common routes are by mouth (oral dosing), by application to the skin (topical, percutaneous), by injection into a vein (intravenous, i.v.), by injection under the skin (subcutaneous, s.c.), by injection into the abdominal body cavity (intraperitoneal, i.p.), by injection into a muscle mass (intramuscular, i.m.) or by inhalation. Accidental intake of environmental chemicals is usually by inhalation, by skin contact or by mouth. The latter intake includes direct uptake by mouth (usually by young children) and indirect movement of inhaled larger particles which are deposited in the nose and swallowed after being moved to the mouth cavity. The proportion of toxic substance that reaches the target organ within the body, and the time after administration at which it gets there, both vary with the route taken. Thus intravenous injection usually produces the quickest effects with the lowest dose. Of the routes of accidental intake, inhalation is usually fastest, followed by mouth or skin. However, the chemical nature of the particular substance controls the way it enters the body (see Chapter 5).

More complex units of dosage may be necessary for some routes. Thus for inhalation, to express the total amount of chemical received by the body, it is usual to speak of a certain concentration of the substance breathed for a unit time. This product could for example be milligrams per cubic metre per minute ($mg/m^3/min$) and is called the Ct value. A Ct value which causes death of 50% of a population then becomes a LCt_{50} value. Such an expression is understood to refer to a set rate of breathing. If the subject is breathing more rapidly, more of the chemical will be received into the lungs; if breathing is stopped (breath-holding is possible for about 15 s) the subject will receive much less. However, if you now try holding your breath you may observe slight air movements in and out, so that although you may think you can hold your breath longer than 15 seconds, in fact you have not stopped breathing entirely. Practised swimmers are best at breath-holding but even they cannot hold completely, so don't rely on this to get you out of a smoke-filled room or a toxic cloud.

Limits for occupational exposure are more conveniently based on an 8 h workshift, with 16 h or more of recovery for removal of the chemical from the body and/or breakdown to less toxic products. Such a maximum allowable concentration is usually given in parts per million (ppm) or milligrams per cubic metre (mg/m^3).

Dosages for chronic exposure become difficult and complex to express. If the intent is to mimic the long-term exposure to a chemical that acts through the skin, then single daily applications of a stated amount may be made to the skin of the test animal for a long period. In the real case of chronic exposure of humans to a chemical, the dose is very difficult to estimate; the concentration of the chemical has very likely fluctuated widely over the time of exposure and only broad estimates can be made of what the subject has been exposed to.

We have now accumulated some qualifications to put on the dose concept. Mass of chemical plus indication of purity plus size of animal or human subject plus route of administration plus time of exposure (if not a single dose) help to define more closely what causes the observed effect. It may be helpful to quote some sample figures to give you a feel for them. Table 4.1 gives some values for substances selected at random. There are obvious deficiencies; many figures are reasoned guesses, especially those relating to man. Such values are obtained from reports of suicides or accidental poisonings (e.g. for the insecticides and hydrogen cyanide) or from studies on patients who have come to their doctors with a specific

Table 4.1. *An illustrative assortment of toxicity data for a variety of chemicals*

Substance	Effect	Mass, Concentration etc.	Subject, size	Route	Duration of exposure
Ammonia	Eye injury	700 ppm	Assume adult human	Direct exposure of eye	Not given
Ammonia	None (maximum allowable concentration)	25 ppm	Adult	Mainly inhalation	8 h workshift
Hydrogen cyanide	Death	$5\,g/m^3/min$	Adult	Inhalation	Expressed in C_t value
Sevin (insecticidal carbamate ester)	Death	1.4 g/kg	Rat	Subcutaneous injection in lard	Single dose
Aspirin	Death	About 25–30 g	Adult human	Oral	Single dose
Aspirin	Death	200 mg/kg	Rat	Oral	Single dose
Aspirin with phenacetin	Gross damage to kidneys	10 tablets per day	Adult human	Oral	Average of 14 years
Malathion (insecticidal phosphate ester)	Death	About 1 g/kg	Adult human	Oral	Single dose
Physostigmine (from calabar bean)	Death	1.2 mg	Adult human	Intravenous	Single dose

complaint (e.g. for chronic aspirin plus phenacetin consumers). More accurate data are obtainable from animals, but their value in human terms is debatable. It is very difficult to get any quantitative data at all for those chemicals to which people are constantly exposed, which have no obvious short-term toxicity, but are suspected of having long-term effects. Precise toxicity data are thus obtainable for those chemicals and those animals which are of little social consequence; the socially most significant data are the most elusive.

This leads to another vexed and complex question. How do you extend data obtained on animals to man? There is no easy answer; the arguments about this continue and are different for the various classes of chemicals discussed. You can take a lethal dose value obtained on rats, express it in milligrams per kilogram, multiply by an adult human body weight and arrive at a figure in milligrams which may or may not have any meaning. Rodents are the most convenient laboratory animals to use, but are biologically unlike man, whereas monkeys are expensive and inconvenient to use but probably a better model. I cannot make any generalisations here, as each case seems to be a matter of dispute to be settled on the particular evidence available. Differences between species cause problems, but individual differences within one species are also obvious. We have all observed a sensitivity of some persons to a chemical which others tolerate. This may be to a complex chemical of biological origin (such as pollen causing hay fever) or a synthetic product (detergent causing a skin rash). When reporting the toxicity of a chemical, it is necessary to state the species upon which the observations were made.

Since populations of individuals do vary as noted above, the dose for an effect is usually related to the quantity of chemical producing the effect in 50% of the persons (or animals) observed. Terms such as LD_{50} are used; in this case it means the quantity of the substance required to kill 50% of the group of animals exposed to the chemical. Death may not be the endpoint observed. If you are examining a group of people to determine the cause of a dermatitis, you may report the quantity of chromate found to produce a rash on 50% of them as an ED_{50}, i.e. effective dose for 50% of the population. The accuracy of determination of the LD_{50} or ED_{50} increases with the size of the population examined. The particular significance to be attached to an evaluation of a dose is estimated by statistical methods, which play an important part in the interpretation of such data. This is particularly so in the consideration of environmental data, which are gathered from scattered and haphazard events of low frequency. I do not propose to consider statistical sampling and analysis methods here; it is a complex subject for the layman and is covered in several texts for the technical person. I wish to make the point, however, that statistical analysis cannot transform doubt into certainty; it can only analyse and categorise the degree of doubt. If the observations show uncertainty, then no analysis will remove that uncertainty. In other words, there is no certainty. All statements are graded with the probability that they are true, so that it is highly probable that thalidomide is linked to birth defects, less probable but still fairly certain that smoking increases the risk of lung cancer, and highly improbable that tea-drinking is a cause of baldness.

The basic message of this chapter is that the term 'poisonous' is meaningless on its own. It assumes more and more meaning as the conditions under which the chemical is harmful are more clearly and fully stated. I was surprised the other day to open a book entitled *Toxicology of Drugs and Chemicals* and find that it fell open on an entry for 'milk' which occupied a page. Apparently if you insist on drinking a gallon of milk a day you will end up with excessive calcification, causing pain in your joints.

It is instructive to reverse the usual line of reasoning and consider the beneficial effects of chemicals. The metals zinc and copper (among others) are found in the body as parts of enzyme systems; they are therefore essential for normal function, not just beneficial. Yet these two metals are commonly regarded, quite rightly, as being toxic and dangerous. The explanation of this paradox is that the quantities required to maintain the essential levels in the body are very small, maybe micrograms per day, whereas the toxic levels are tens of thousands of times higher, in the order of grams for a single dose. This concept is illustrated in Fig. 4.1, in which the chemical in question could be a metal (as above) or a vitamin. The dose range shown is just for illustrative purposes, it will vary for different chemicals. At very low doses the animal or organism is harmed because it is not getting enough of an essential material. As the dose increases, there is an increase in benefit to the animal (zone B) as the essential need is satisfied. Then very slowly an excess produces a harmful effect, so there is a long transition zone (C) in which there is no obvious, dramatic effect of the chemical. This transition blends into the final zone (D) in which the harmful results of high doses are apparent. Is this chemical toxic, or is it beneficial? It is both.

Many chemicals show this transition from beneficial or essential effects at low levels of exposure to toxic effects at higher levels. There is thus a continuity between benefit and detriment; the aim of the toxicologist is to define the safe region between.

Therefore toxicity is not a 'yes or no' property. There are some chemicals which are so very toxic at low doses that they score a near black rating. Others are so innocuous as to be nearly lily white. However, pure black and pure white do not occur. We are looking at various shades of grey. Toxicity is a continuum on this grey scale and, given adequate data, we can place a chemical with reasonable accuracy on the scale but we cannot and must not make absolute statements on toxicity.

To conclude this initial consideration of toxicity, we may ask if there is any simple definition of toxicity levels, i.e. does 'high' or 'low' toxicity have any defined meaning? Unfortunately, I think not, although attempts have been made to class chemicals in this way. Table 4.2 is derived from a number of sources and attempts to illustrate the sort of classifications that can be made, with examples of the classes of chemicals involved. The toxicity levels blend into one another, so that it is a very artificial exercise to try and distinguish them. In effect, quantitative data must be given with any statement about toxicity, and there is no good shorthand way of expressing this.

It is worth pointing out briefly here that a lot of the usefulness of chemicals as pesticides and antibiotics depends on their varying effects on different creatures, which we gave above as being a problem in translating animal toxicity data to humans. This concept of selective toxicity is most critical to the design of antibiotics, as the human patient takes the drug

Fig. 4.1 Beneficial and harmful effects are exerted on a living organism by one chemical at different dose levels. The diagram shows transitions between benefit and harm as the dose increases. The effects could be normal growth versus stunting or death, or fertility versus sterility, or many other properties. Note that the dose axis has a logarithmic scale. The dose units are assigned arbitrarily.

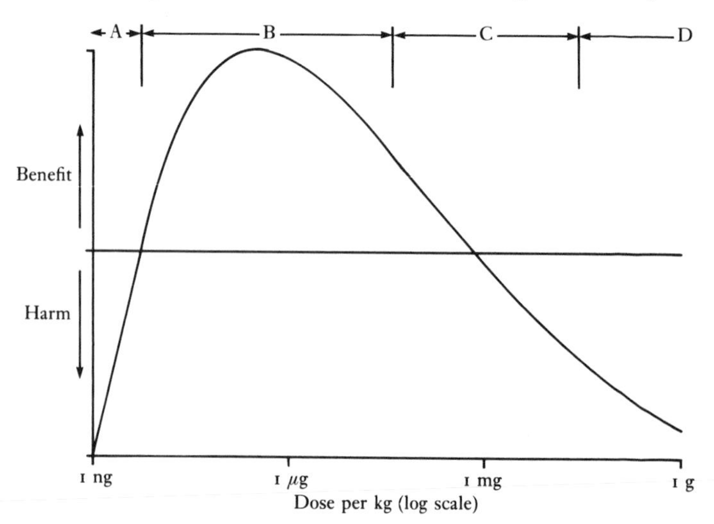

within his body. Thus penicillin is useful as it will kill bacteria at concentration levels that do no harm to the cells of the human body. Higher concentration levels may well do so. Selective toxicity is a less pressing requirement for pesticides as we need not expose ourselves to them, although it is most desirable. For example, we need to handle concentrates of herbicides or insecticides with care, and make sure their residues do not contaminate food to a dangerous degree. Various classes of pesticide differ within themselves as to their human toxicity; among herbicides 2,4-D and 2,4,5-T have much lower acute human toxicity than paraquat, and are therefore much safer to handle. This concept of selective toxicity is something to be kept in mind when other aspects of chemical toxicity are explored.

In order to understand more fully the concept of toxicity, it is necessary to know how chemicals are tested for toxicity. I have summarised the principal methods in Table 4.3, and further details may be found in references 1 and 2. Acute lethality is the crudest test, but some information can be gained by pathological examination of the organs of the animals after death. The determination of the LD_{50} is necessary if you need a quantitative figure for the toxicity of a chemical for regulatory or other purposes and, because it is quantitative, it is superficially attractive to the

Table 4.2. *Levels of toxicity of various classes of chemicals*[a]

Toxicity Class	Lethal dose[b]	Examples of chemicals in the class
Biotoxins	Much less than 10 μg/kg	Ricin (from castor oil plant), botulinum toxin
Supertoxic	10 μg/kg–1 mg/kg	Nerve agents. Some carbamate esters. Atropine (from deadly nightshade)
Highly toxic	1–50 mg/kg	Some organophosphorus insecticides, sodium cyanide, vitamin D, T2 mycotoxin
Moderately toxic	50–500 mg/kg	Organophosphorus and organochlorine insecticides, barbiturates
Slightly toxic	0.5–5 g/kg	Aspirin, many commercial ‘solvents’
Hardly toxic (considered harmless)	Greater than 5 g/kg	Not too many synthetic chemicals would fall in this class

[a] Toxicity is here expressed as death after acute poisoning.
[b] The dose given is for one administration by mouth or by intravenous injection. For some chemicals, the two routes may give quite different effects.

Table 4.3. *Summary of test methods for the toxicity of chemicals*

Test	Procedure
Acute lethality (LD_{50})	Treat animals with a range of doses, and count number of dead after a set time (usually 24 h). Make pathological examinations of dead and survivors. Usually simple dosing route (e.g. intravenous).
Whole-animal monitoring	Lightly anaesthetise animal, then apply instruments to it to measure physiological parameters, such as respiration rate, heart rate, arterial and venous blood pressures, electrical activity of brain and heart, etc. Increase dose of chemical very slowly (intraveneous infusion is preferred) and follow changes in the parameters being measured.
Chronic toxicity	Repeated treatment of animals with sublethal doses. Do pathological examinations at set times (1 month, 6 months, 2 years, for example) and note all unusual features. Route of dosing must reflect practical risk (e.g. oral for foodstuffs, inhalation for vapours).
Carcinogenicity	As for chronic toxicity, paying particular attention to tumours and incipient changes in tissues. Use cofactors to increase artificially the production of cancers (e.g. croton oil).
Mutagenicity	Test ability of chemical to induce growth in a microorganism that has been inhibited by a genetic manipulation (Ames test). Pre-incubate chemical with enzymes from liver, to see if it is readily transformed in the body to a mutagen. Check for any gross effects on the structure of nucleic acids in chromosomes, by microscopy.
Teratogenicity	Treat pregnant mother with chemical at appropriate doses by the relevant route. Choose period of pregnancy most sensitive to chemicals (first third of incubation period). Count number of aborted foetuses, stillbirths and survivors. Make pathological examination of all foetuses.

more mathematically minded researchers. However, it gives very little real information about the mechanism by which the chemical exerts its toxic effect. Whole-animal monitoring is much better for this; it is possible to detect the sequence of events that leads to death, and thus help locate the body system that is first disrupted by the chemical. Chronic toxicity testing, with carcinogenicity and teratogenicity, is a tedious business which requires many animals and much time. It is very desirable to have shorter methods but, at present, the bacterial mutation tests for mutagenicity are all we have (see also Chapter 15).

There are two distinct aims in testing for toxicity. One is to satisfy a requirement that a chemical be tested in a defined way and that certain parameters be measured. The second aim is to increase our understanding of why chemicals are toxic and to describe fundamental mechanisms of toxicity. If the first aim is pursued in a routine way purely to satisfy the limited requirement, then a great deal of information helpful to the second aim is lost. Testing against a requirement absorbs much more effort than research, so we need to encourage a more inquisitive attitude in routine testing.

We have seen early in this chapter that toxicity must be related to dose, i.e. that a biological response is related to the dose given. This is a subject of some importance in discussions of the effects of chemicals, together with the related question of what happens at a dose level below that which produces an observable response. Therefore we will discuss the dose–response relationship further.

In very many cases an increase in dose of a chemical produces an increase in the biological response to that chemical, whether the response be toxic or beneficial (e.g. vitamins). Further, the relationship between dose and the consequent response is often simple. Thus if we look at a relatively narrow range of concentrations between zones A and B (Fig. 4.1), then we will see an increase in benefit with increasing dose. Similarly, in zone D we will see an increase in toxicity with increasing dose. Over the range of zone B, we encounter a much more complex relationship between dose and response. It is quite routine to check the dose–response relationship when embarking on the examination of the toxicity or the pharmacology of a chemical. To reassure oneself that the experimental method you have set up is good, you can check the response of the system to increasing doses of a chemical with which you are familiar. If the relationship between the two conforms to past, satisfactory experience, then you can go on to examine novel chemicals. If the relationship is erratic or not clear, then you first examine your experimental procedure for faults.

The relationship between dose and response may be quite complex, and the latter is subject to the individual variations found in any population of animals. Figure 4.2 shows the kind of variation around a central dose that is often found. The same information can be replotted as in Figure 4.3, and from such a curve it is easy to estimate the LD_{50}. It is mathematically desirable to process the data further to give a straight-line graph, and for

Fig. 4.2 The dosage of a chemical necessary to cause death in a test animal. A population of 53 animals was used. The response of individual animals varies about a central value (approximately 10 mg/kg).

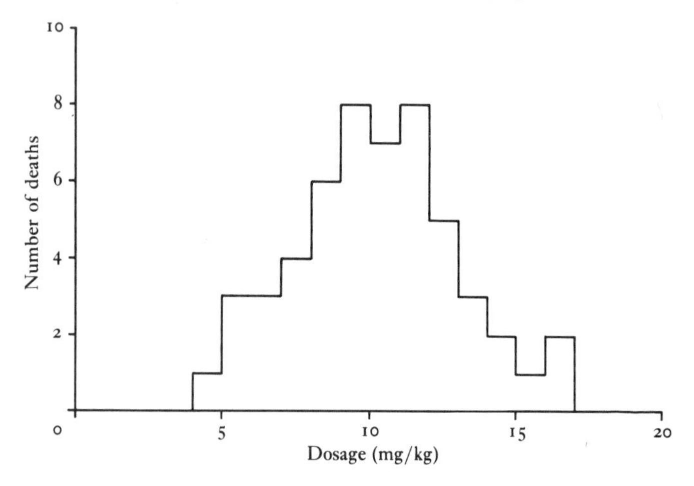

Fig. 4.3 The same data as in Fig. 4.2, plotted to produce a continuous curve of frequency distribution. An estimation of LD_{50} is now relatively easy.

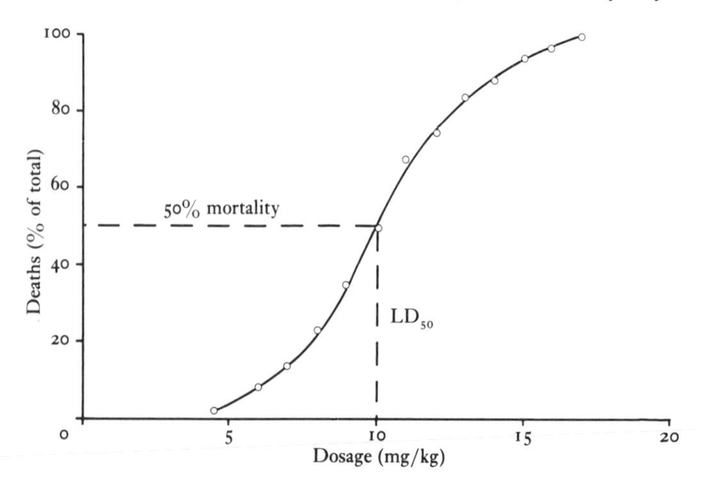

many dose–response relationships this can be done, as in Fig. 4.4. The vertical axis is converted to probability units (or probits) and the dosage to a logarithmic scale. In crude terms, the first action takes account of population variations, and the second of complex relations between dose and response. The details need not concern us now; the point to remember is that the quantitative relationship between cause and effect is often not simple.

There have been reported instances in which lowering the dose has increased the effect (3). This is not inconceivable if you are working in a concentration zone between beneficial and harmful effects. It is also a principle of homeopathic treatment (Chapter 9). However, if this anomalous relationship is observed, the first thing to do is to check your method to exclude experimental artefacts because, in 99% of cases, the response increases with the dose.

What happens at low doses, below that at which obvious effects start to show? In order to understand this question, we have to appreciate the complexity of biological systems. The response that we observe (sickness, nausea, chronic degeneration of an organ, death, etc.) is not a primary response. It is the end result of a very complex set of events, in which there may be very many steps between the actual effect of the chemical and the response that is observable by us. For example, we can examine the inflammatory response of the skin to a liquid which kills living cells, such as

Fig. 4.4. The curve of Fig. 4.3 has been transformed to a straight line by plotting the logarithm of the dosage, and using a probit scale for the response. This is a common form of dose–response graph, and can be used in pharmacological tests other than lethality estimates.

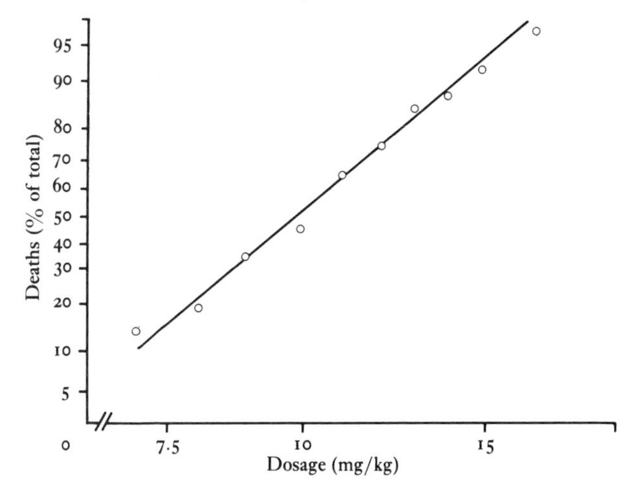

mustard 'gas' or a solution of T-2 toxin. The primary event is the death of the rapidly growing cells which replace the outer epidermis as it is worn off. Heat or ionising radiation will have the same effect. We do not see the death of the cells; we cannot. What we do see is reddening due to increased blood supply, following by blanching and swelling as the skin tissue loses control of liquid balance and liquid enters the damaged area (oedema). Then there may be blistering as the outer epidermis is pushed away from the remaining, deeper skin. The liquid in the tissue is forced to the surface and exudes, drying out to form a protective layer (scab). Then the wound slowly heals as the living cells on the margin of the area grow back into it below the scab. This process is quite familiar to you as the healing of a thermal burn. The point is that the primary event that triggers this process is not obvious, and we still do not know exactly how all the events are linked. We can deduce that attack by the toxic chemicals on the multiplying cells is the prime event, but we cannot be sure that they (mustard gas, T-2 toxin) do not trigger other elements of the inflammatory response.

Death is a more complex event again. Ultimately the victim dies of respiratory failure or heart failure, but we only say that because they are the most obvious events. Was the respiratory failure due to paralysis of the muscles which ventilate the chest, or to failure of the nervous signals from the brain to those muscles? If so, are the nerves or the brain affected? Or has the blood supply to the brain been interrupted? The questions go on and on.

The lesson for us is that toxic chemicals at low dose may exert an effect on the biological system which we cannot see by our crude observation of endpoints. Are those effects detrimental in a subtle or long-term way? If they are what is a safe dose?

Organophosphate esters are used as insecticides and (unfortunately) as chemical warfare agents. They cause death by interfering with the transmission of nerve signals to muscles by inhibiting the enzyme cholinesterase. The organophosphates used as insecticides are less toxic to man than to insects, but deaths have occurred by suicidal drinking or, in a few cases, by gross carelessness in use. Spray operators or persons handling sprayed produce fresh from the fields (e.g. tobacco workers) may absorb small quantities of the insecticides. The cholinesterase in their blood will then show a depressed activity, but there will be no other sign of poisoning. Exactly the same is true of smokers or those of us who drive in city traffic; our blood haemoglobin is partly converted to the useless carboxyhaemoglobin by the carbon monoxide in cigarette inhalations or vehicle exhausts, but no obvious sign of poisoning is seen. Again, exposure to low concentrations of hydrogen cyanide poisons some of our cytochrome oxidase, but symptoms do not show.

In this form of low exposure to toxic agents, the body has available reserves of the target system, so that the system in normal operation is not affected. However, it may be in stressful situations (e.g. we want all the oxygen we can get as we run for the train). The signs of toxicity are also indications of imminent trouble. The answer to our original question is therefore that such effects may be detrimental in some situations and that the low doses at which they occur are on the margin of danger. A safe dose must be lower; how much lower is a guess only.

However, carcinogenic substances present a different problem. It can be argued that one molecule of a carcinogen can cause a change in a nucleic acid, which may result in an altered gene which, in the right environment and with the right cofactors, will result in a cancer. Therefore no concentration level of a carcinogen can be entirely safe. This is true. But one chemical change which occurs among all the other changes caused by natural and artificial means (radiation and chemicals) is not going to add much to the risk of the individual getting a cancer. The risk is statistical; at very low doses the risk is very, very low, so low as to be negligible.

There are no absolutes in the study of toxicity; all risks are statistical, relating to the dose of chemical received. The understanding of this chapter is fundamental to the rest of the book, all other arguments are of necessity built on it. I have discussed the interpretation of safe levels further in chapter 15, and a clear exposition of the matters discussed in this chapter is given in reference 4.

If we are to understand the significance of chemicals in our world, we must stop seeing toxicity as an absolute, and appreciate the shades of grey. We must also appreciate the statistical nature of risk, but above all we must cast aside any tendency to superficial generalisations and examine each question of toxicity freshly and with reference to all the information available.

References

1. Paget, G. E. (ed., 1970). *Methods in Toxicology*. Oxford: Blackwell.
2. Garrod, J. W. (ed., 1981). *Testing for Toxicity*. London: Taylor & Francis.
3. Kon, S. H. (1978). Underestimation of chronic toxicities of food additives and chemicals: the bias of a phantom rule. *Medical Hypotheses*, **4**, 324–39.
4. Pascoe, D. (1983). Toxicology. *Studies in Biology*, no. 149. London: Edward Arnold.

5

Entry and exit: how chemicals get into the body, and how they are eliminated

A toxic chemical on the outside of the body has to be moved to the inside in contact with living tissue to exert its effect. The location of the target in the body varies depending on the nature of the chemical but, once the chemical reaches the bloodstream, it is carried round the body very rapidly so that in most, but not all cases, it is effectively then in contact with the target organ. The difference in apparent toxicity of a chemical when administered by different routes was mentioned in Chapter 4. A comparison of the effective dose by mouth, inhalation or skin permeation to the dose by intravenous injection allows a rough assessment of which route poses particular barriers to a chemical.

The three natural routes of entry are through the gut, the lungs and the skin. The gut and the lungs each have surface areas of lining much greater than might at first be apparent: in the first this is due to finger-like protrusions (villi) into the cavity of the gut; in the lungs it is due to the many fine chambers (alveoli) at the ends of the air passages. These increase the area of contact enormously and help the passage of chemicals into the body. The skin of an adult human has a superficial area of approximately 1.8m^2,

Fig. 5.1 A diagram of the principal modes of chemical entry and exit, to and from a simplified animal body. Movement within the body is by circulation in the blood (lymph plays some part in this).

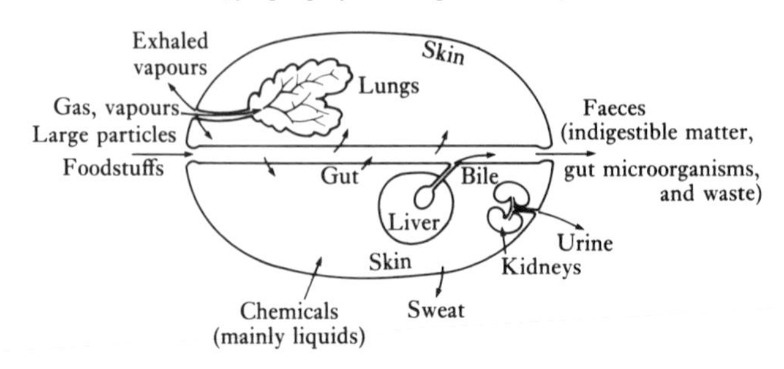

but folds and minute irregularities increase this greatly. The modes of entry of chemicals to the body are illustrated very diagrammatically in Figure 5.1; we will now consider these in more detail.

From studies on embryology and on the development of primitive animals it can be seen that the cavity of the gut is part of the environment that has been pushed into the main body of the animal to form the continuous tube of gullet, stomach and intestines. This cavity is partly sealed off from the main environment by the mouth and anus, and further it is carefully controlled by the body. At the upper end of the intestinal tract (Fig. 5.2) are added digestive juices, salts and water, if necessary, to produce a localised environment suitable for the digestion of foodstuffs. Towards the end of the tract all the useful substances are reabsorbed together with the digested foods, so that the body conserves its supplies of material. The fate of a foreign chemical that gets into the gut is quite variable, depending on the particular type of chemical. Movement of a

Fig. 5.2 Functional diagram of the intestinal tract showing the main regions and their function in absorbing food and excreting waste.

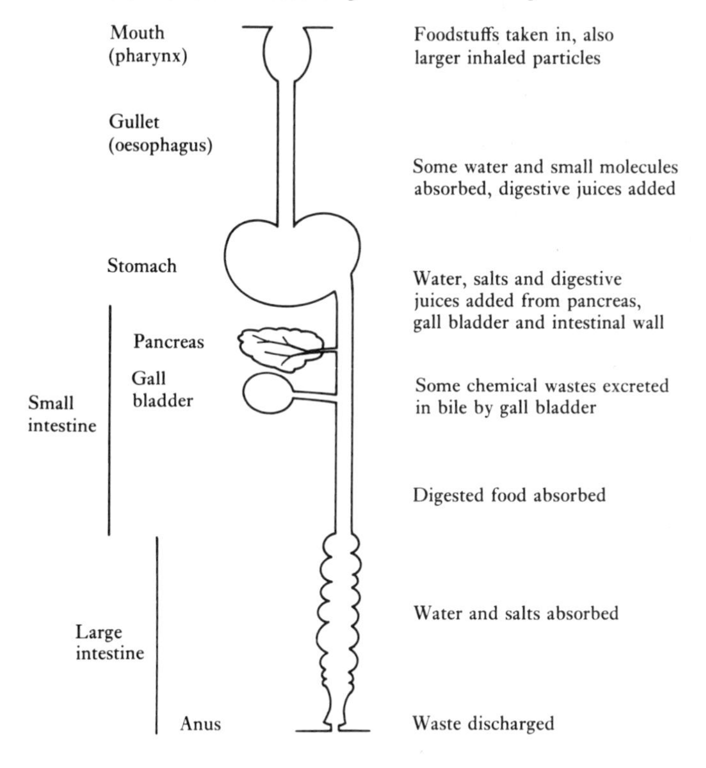

chemical across the gut wall into the bloodstream may be by passive diffusion or it may be assisted by a carrier mechanism in the cells of the gut lining. In the latter case, the cells may be able to move certain chemicals against a concentration gradient by the expenditure of energy. Such a process requires a complex cellular mechanism involving a chemical carrier system coupled to a source of energy. Of course the carrier mechanisms are not there to carry strange chemicals into the body; they normally act to help the body secure particular nutrients from the foodstuffs.

Chemicals which exist as small molecules, such as alcohol or cyanides, are rapidly absorbed from the stomach and gullet, whereas larger molecules may be more slowly absorbed from the intestines. Chemicals which are soluble in both water and fats are usually taken in more rapidly than those which show solubility only in one or the other. Such generalisations are, however, rather dangerous to make since there are many exceptions. Thus, some chemicals that are of low molecular size have a positive electrical charge on them which tends to slow absorption. Therefore these positively charged compounds pass through the intestinal lining at a very slow but steady rate. One result of this is that the proportion of the dose of a drug of this type that is absorbed by the gut depends on the movements of the gut and the rate of passage of foodstuffs through it. Since this varies from person to person, it makes the required dosage very difficult for the physician to estimate.

Microorganisms are normal inhabitants of the intestinal tract, particularly at the lower end. Thus a fair proportion of faecal matter originates from microorganisms. Chemicals taken into the gut may be modified by this intestinal flora, so that the consequent products could either be less easily, or more easily absorbed into the body than the parent chemical.

The gut is a very efficient system for absorption of materials, so it follows that any foreign chemical that gets into the mouth will find its way sooner or later into the body (in its original or in an altered form). The variable is the time required; advantage may be taken of a slow absorption rate to remove a toxic chemical by inducing vomiting or washing out the stomach. This statement must not be interpreted as saying that the intestine is not a barrier between the body and the environment. It is. It is not a complete barrier and its efficiency varies widely depending on the chemical seeking access.

The lungs are designed for the efficient exchange of gases between the air and the bloodstream, oxygen inwards and carbon dioxide outwards. Air is drawn in through the nose (or mouth, or both) down the windpipe (trachea) and through successively branching smaller tubes to the alveoli at the end (Fig. 5.3). During its passage deep into the lung, the newly inspired air is mixed with air remaining in the dead spaces from the previous breath. Oxygen from the air eventually is absorbed into the blood through the lining of the alveoli and passages, and carbon dioxide diffuses from the blood into the air. The exhaled breath thus contains excess carbon dioxide and water vapour over the ambient air, but less oxygen. The frequency and depth of breathing is quite variable, so that the quantity of a chemical vapour inhaled in a certain time also varies. Frequency can range from about 14 breaths per minute for an adult at rest to 25 or 30 for someone doing hard work, and the volume of air taken in per breath (tidal volume) can vary from 0.5 litre (at rest) to 1.0 litre (working). Therefore the quantity of air taken in per minute is very dependent on the subject's mode of breathing, which of course influences the apparent toxicity of a vapour when expressed as LCt_{50} in $mg/m^3/min$ (see Chapter 4). The picture is further complicated by the fact that the respiratory system is not completely emptied and filled at each breath; there is a dead volume of air with which the newly inspired air mixes.

The alveoli at the end of the bronchial tree should not be thought of as

Fig. 5.3 Functional diagram of the respiratory tract to illustrate the main regions and their function in gas absorption and particle collection.

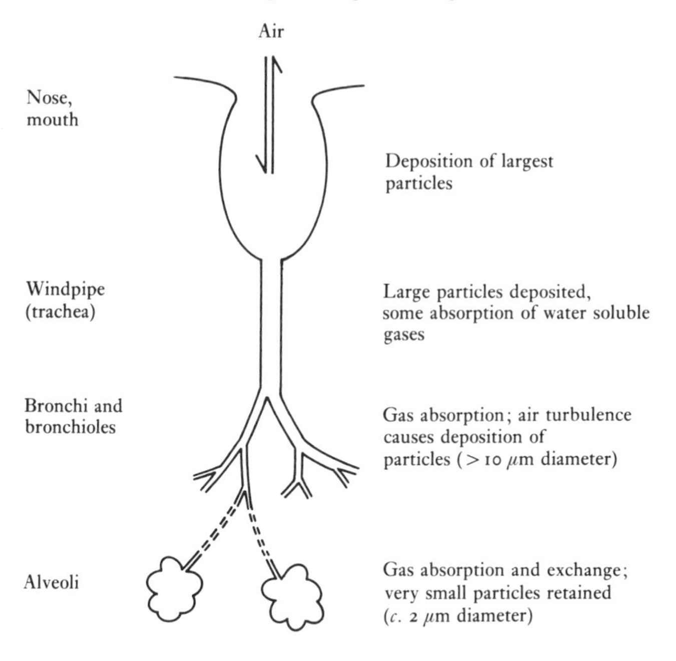

the only places where gases and vapours are absorbed by the respiratory tract. In fact a vapour which is very soluble in water probably would never reach the alveoli. In its passage down the narrow tubes of the lung it will contact the damp walls, be dissolved in water and carried into the body fluids. Thus hydrogen cyanide and some insecticide vapours are probably absorbed largely in the nose and coarser tubes of the bronchial tree but, as their effects extend over the whole body once they are absorbed, this fact is of little practical importance. It is important for those vapours which directly damage the lung tissue, such as chlorine, acid vapours (hydrochloric acid) and ammonia.

Chemicals which are not readily soluble in water may be less rapidly absorbed through the lungs but, because of the great efficiency of the lungs (due to the enormous effective surface area), it must be assumed that all chemical vapours are capable of entering the body through the respiratory tract.

Solid particles and liquid aerosols can also enter the respiratory tract, and some can cause serious toxicological problems (witness the tragic effects of breathing asbestos particles). Such particles are deposited in the respiratory system at a location which depends on their size. Thus larger particles (greater than 10 μm in diameter) are deposited in the nose, and are subsequently lost by sneezing, by blowing the nose or by swallowing. The latter is important, since the route of entry to the body may then be by the intestine, even though the particles were inhaled. Large particles of solid or liquid lodged in the nose and which are very water soluble may be absorbed through the lining of the nose. The importance of the nose as a trap for dust is always apparent to me when I finish a dusty job such as clearing out the garage and then blow my nose, to see a white handkerchief turn grey. Further filtration occurs as the air is carried down towards the alveoli. There are approximately 20 forks of the airways between the windpipe and alveoli. This branching causes turbulence in the airstream and any particles in the air have a good chance of being thrown onto the moist walls of the airways and trapped there. Particles smaller than 10 μm can pass into the finer passages of the respiratory tract, so that those of 1–2 μm diameter are deposited in the alveoli. Below this size, the efficiency of deposition falls again, so that very small particles are exhaled again. Of the 1–2 μm sized particles, about 50% may be retained in the alveoli and the remainder exhaled also. The figures are for spherical particles; if the material is a fibre, the diameter is still the main factor, so that quite long fibres may enter the alveoli.

The lungs are therefore protected to some degree by a filtering system formed by the nose and the airways. This filter protects the alveoli of the lungs from the more water soluble chemical vapours and from solid and liquid particles greater than 10 μm in diameter. The smaller particles that can get into the alveoli (spheres about 1–2 μm diameter, or fibres much longer but of the same diameter) may cause disease of the lungs known generally as pneumoconiosis, or more specifically as asbestosis, silicosis, etc. when the particular kind of dust is known. Particles thus mainly affect the lung itself; vapours or gases enter the body through the respiratory system and may cause local or widespread effects.

The third pathway for chemicals to enter the body is through the skin. The study of the movement of chemicals through the skin has been pursued erratically in the past, but today there is recognition of the importance of the skin and an awareness of our lack of knowledge concerning it. The skin is far from being a passive sack surrounding the body, it has interesting properties which control the movement of chemicals across it. In the small, primitive aquatic animal the skin was a principal site of the exchange of nutrients for waste products from the environment to the body. Large terrestrial animals such as humans have developed special organs (lungs, guts, kidneys etc) for these functions, so that the importance of the skin has perhaps declined, but remains significant.

The outer skin (epidermis) is constantly growing from below, pushing outwards new cells which become loaded with a protein, keratin. These cells become flattened and die off, ending up as flat plates of keratin on the outer surface. The thickness of this horny layer (the stratum corneum, see Fig. 5.4) varies at different parts of the body surface, e.g. very thick on the soles of the feet, thin on the skin of the face. Below the epidermis is the dermis, a thicker tissue which contains blood vessels and performs all the maintenance functions for the skin. The barrier to the diffusion of liquids and dissolved chemicals across the skin is the horny layer of keratin, so that the properties of this are very important. Some chemicals can enter the body through the pores created by the sweat ducts and hair shafts but, in terms of quantity transported, these pores are not important as they form 1–2% of the skin surface. You may have heard it said that chemicals are more readily absorbed by hot skin 'because the pores are open'. This is nonsense. There are no pores which open or shut to allow chemicals to enter or not. If hot skin does absorb more of a particular chemical, it is due to change in the properties of the skin (e.g. hydration), which may, in some cases, be a result of sweating through the sweat ducts, not to the opening of those ducts.

The properties of the horny layer vary with surrounding conditions. In hot, humid atmospheres the keratin can absorb much water, especially when the person is sweating, and the thickness of the horny layer is increased. Conversely, in warm, dry air the keratin contains little water. The barrier properties of the horny layer to a particular chemical may be altered with change in water content. The temperature of the skin of a person in a comfortable environment is about 32 °C, i.e. 5 °C less than deep core temperature. The skin temperature will change with extremes of hot or cold in the surrounding air, but not enough to have much effect on the passage of chemicals through it. Since the horny layer is the main barrier, any damage to it will allow chemicals to enter the skin much more readily; persons with cuts or abrasions on their hands should therefore handle chemicals with greater care than normally.

Formerly, the skin was regarded as a barrier to practically everything. Now it is being found that it is permeable to many chemicals, although the rate of entry is slow. Thus it is to be thought of as a delaying barrier, rather than an absolute one. Very roughly, 30 min or more commonly elapses between the application of a chemical to the skin, and the appearance of a significant concentration in the blood. A large dose on the outside (e.g.

Fig. 5.4 Schematic diagrams of (*a*) vertical section through outer skin layers showing main structural features, and (*b*) functional features which control the diffusion of vapours and liquids through skin. The dimensions, which are approximate, are given in micrometres (μm). The thickness of the layers varies widely between different parts of the body surface.

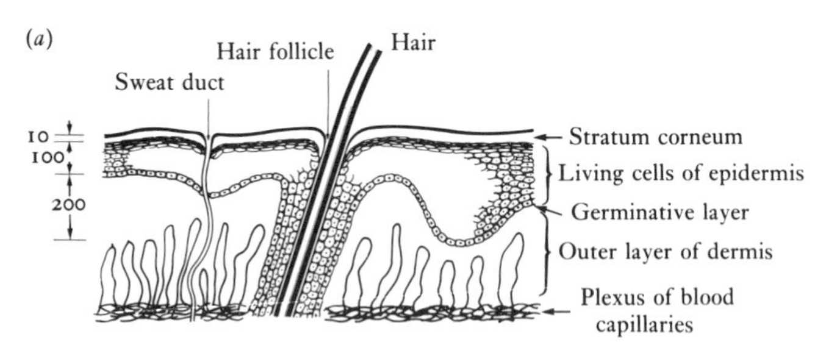

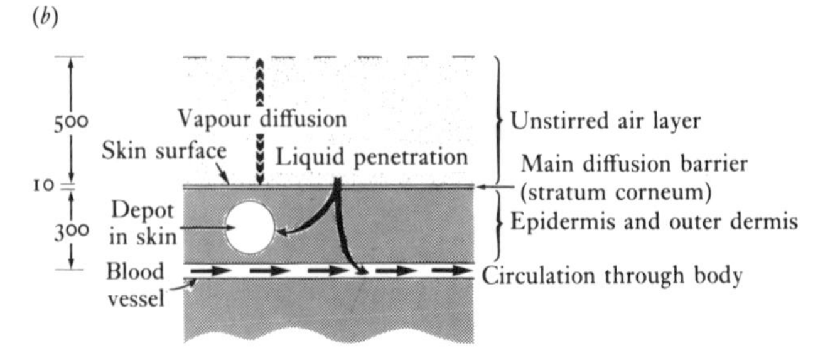

from an extensive spill of a liquid chemical) may, however, appear in the blood and have serious consequences within a few minutes. The skin is most permeable to small molecules of organic chemicals, such as various alcohols and organophosphorus insecticides. The importance of the latter is obvious; contact of insecticide concentrates with the skin should be avoided. It is less permeable to salts and compounds soluble only in water. Solid particles cannot penetrate the skin unless they are water soluble, in which case they will be gradually dissolved by surface moisture and taken in as a solution. The skin is also permeable to oxygen, carbon dioxide and water, so that the normal requirement of the skin for oxygen can partly be met by direct diffusion.

The permeability of the skin is very susceptible to alteration by chemicals, so that a substance that does not enter quickly by itself may be helped in by a second chemical which causes temporary or permanent damage to the barrier properties of the horny layer. An example of a chemical which increases the permeability of the skin to other chemicals is the solvent dimethylsulphoxide (DMSO). This has been used deliberately to increase the rate of uptake of herbicides into plants, which depends on a mechanism similar to that of human skin permeation. A second, unexpected problem occurs because the skin possesses storage properties. A chemical may be taken up by the skin and slowly pass through. At any one time there is therefore a quantity effectively in store in the skin. An event on the outside of the skin (application of a second chemical, change in water content) may cause a rapid release of this stored chemical, resulting in a sudden increase in its concentration in the blood. If the first chemical is toxic, such an increase might be just enough to produce an obvious toxic effect.

Because of the factors mentioned in the paragraph above, it is very difficult to predict how well a chemical will penetrate skin, especially if it is in a mixture with other substances, as is often the case with proprietary preparations. Also methods of cleaning skin need to be examined critically, since washing with solvents (kerosene, alcohol etc.) may cause more harm than good.

The above discussion summarises the main factors which regulate the passage of chemicals across the skin. Once these chemicals enter the dermis and reach the skin capillaries, they are transferred rapidly around the body in the bloodstream, so that all the time delay between contact and effect is due to the passage across the skin. However, the majority of chemical effects due to skin contact are changes in the skin itself, not effects on the body as a whole. Dermatitis is a very common occurrence and may be caused by virtually any of the chemicals created by modern technology.

Chemicals which cause this condition still have to penetrate the horny layer, but then act on the living cells immediately below. The slow onset of chronic dermatitis (due to detergents, etc.) is caused by gradual disruption and breakdown of the horny layer by the chemical, resulting in increasingly easy access to the living cells. In summary, the skin is not an inert barrier, but a delaying system of some complexity which can be penetrated by many chemicals.

Once into the bloodstream from the gut, lungs or skin, a chemical is carried round the body in the blood and can gain access to most organs and tissues of the body. There are preferred routes and particular barriers to movement, which may limit the exposure of some organs to circulating chemicals. There is a functional barrier, the 'blood–brain barrier' which limits the access of particular chemicals to the brain. This presumably is beneficial to the brain, but may also pose problems in the treatment of disease. Thus the specific antidote to organophosphate poisoning (i.e. insecticide or nerve gas intoxication) is a chemical called 2-PAM which cannot cross the blood–brain barrier. It will therefore reverse the effects of poisoning in nerve and muscle outside the brain but not inside. However, in crude terms, it is sufficient to say that access to the blood is access to the whole body.

A toxic chemical which has entered the blood can exert its deleterious effect on the target organ or tissue. Some chemicals interfere with the basic processes of all cells and are therefore toxic to all tissues (except where their access may be limited), others are very specific in their action and interfere with a highly organised complex process. An example of the former is cyanide, which stops the oxidative processes of the cell which yield energy. Organophosphate insecticides and tranquillising drugs are examples of the latter class; organophosphates interrupt the signalling processes of nerve and muscle whereas tranquillisers reduce the complex activities of the brain relating to anxiety and awareness. I cannot list the target organs or tissues of all chemicals, since each of the latter varies slightly in what it attacks. Most of the body is suceptible; the more active organs such as brain, liver, kidneys, often show damage first.

The body has mechanisms for removing foreign chemicals. These mechanisms may remove the strange chemical because it resembles a normal waste product, or a new mechanism may apparently arise, due to the activation of a dormant process by the excess amounts of the foreign chemical. Elimination from the body often involves chemical change. This is followed by an ejection process. The passage out is via the kidneys or sweat glands as urine or sweat, or from the liver as bile through the bile duct to the gut and out (Fig. 5.1). The chemical changes usually result in the chemical becoming more water soluble, which makes it easier for the kidneys to get rid of it. Some chemicals are ejected from the body unchanged; these are usually the very small water soluble molecules. These chemical changes are often described as detoxification processes and, in many cases, they do result in products less toxic than the parent chemical. However, it is dangerous to think of these events as being designed expressly to reduce the toxicity of the chemicals. As far as they are purposive, the chemical changes are for elimination. In some cases the effect may be to increase toxicity. This can have practical usefulness in the design of chemicals for specific purposes. For instance, the chemicals malathion and parathion have a moderate toxicity to man and insects. They are made much more toxic within the bodies of insects by conversion to malaoxon and paraoxon, a conversion which involves the insertion of oxygen in place of sulphur. This process also occurs in man but more slowly, so that the parent compounds appear to be less toxic than in insects. In man the liver has the ability to convert malathion to less toxic products, so that this also contributes to the lower degree of toxicity of these chemicals to man as compared to insects. The liver conversion can be interfered with by other chemicals, so that malathion may unexpectedly appear quite toxic to man in certain conditions of multiple exposure to chemicals. In this way malathion and parathion have a fair degree of selectivity towards insects, and we can employ them as insecticides.

There is no point in attempting to describe the various changes that chemicals can undergo in the body. It is a very complex situation; the pathways of change are still being worked out, and each new chemical brings new research problems. The most practical aspect of the situation is the time factor, which can dictate whether a dose of a chemical is harmful or innocuous. Suppose we have a chemical 'Grot' which is highly toxic to a body organ at a concentration of x mg/l. A man has Grot splashed on his skin and it is estimated that $200x$ mg is spread over the skin. If the chemical became equally mixed with all his body fluids, the concentration would be $3x$ mg/l, assuming he weighed 70 kg and all his body weight was fluid. In fact, all his body is not fluid, so that the concentration would be greater than $3x$ mg/l (a very toxic dose). In actuality, instantaneous mixing will not occur. The chemical will penetrate slowly through the skin. During this time some will be lost from the skin surface by evaporation or mechanical removal, and more will be decomposed as it passes through the skin. The remainder will appear in the bloodstream, not at one time but as a slow trickle. The blood concentration will rise, but as it does so it will start to lose Grot from decomposition and by elimination processes such as excretion in the urine. If the latter processes are fast, Grot will be lost as soon as it appears in the blood, so that it never reaches the concentration of x mg/l of Grot which is toxic to the target organ.

The converse situation may occur. If the person accidentally swallows $200x$ mg of Grot and it happens to be a chemical which is rapidly absorbed from the gut, then the blood concentration will rise rapidly. Further, if the Grot is stable to decomposition and not excreted readily by the kidneys, then the blood concentration will continue to rise to exceed x mg/l by many times, causing drastic effects to the target organ.

This question of rate of transfer from place to place in the body, the balance of rate of accumulation versus rates of loss by various processes, controls the effect which a chemical will have and modifies its intrinsic toxicity considerably. This topic, which was touched on in Chapter 4, constitutes the study of pharmacokinetics. It is obviously of importance in drug design, and it is now required that the pharmacokinetics of new drugs must be studied before they will be accepted by the health regulatory authorities.

What I have attempted here is not to describe the entry and exit of chemicals in precise detail (which would be an impossible task in one chapter), but to illustrate the practical consequences to people of the various ways chemicals may enter their bodies. We are, to a great extent, insulated from the environment but our requirements for energy and growth mean that we have to have mechanisms by which chemicals are exchanged with it. Through these systems foreign chemicals may slip in, unnoticed until some deleterious effect calls attention to them.

Further reading for Chapter 5

I have not found any good general text on the movement of chemicals in and out of the body. The information can partly be found in textbooks of medical physiology, but these tend to gloss over the chemical aspects. Adrien Albert (1) gives a short account of the chemical side, together with an introduction to pharmacokinetics. Accounts of the latter will also be found in textbooks on pharmacology, such as the one by Bacq (2). The permeability of skin is described by Tregear (3) and in the review (4). Deposition of material in the lung is covered by conference proceedings such as those edited by Davies (5), but these are quite technical. Altogether, the subject of this chapter is in need of a good, clear exposition for the layman and non-specialist.

References

1. Albert, A. (1965). *Selective Toxicity*, London: Methuen.
2. Bacq, Z. M. (ed., 1971). *Fundamentals of Biochemical Pharmacology*. Oxford: Pergamon.
3. Tregear, R. T. (1966). *Physical Functions of Skin*, London: Academic Press.
4. Scheuplein, R. J. and Blank, I. H. (1971). Permeability of the skin. *Physiological Reviews*, **51**, 702–47.
5. Davies, C. N. (ed., 1961). *Inhaled Particles and Vapours*, Oxford: Pergamon.

6

Analysis: is it there or isn't it?

The preceding chapter on toxicity and dosage points out the necessity of knowing how much of a chemical is present if we are to predict some toxic outcome. This requires analysis to establish that the chemical in question is indeed there as recognised by its own particular properties (qualitative), and then an estimate of how much of the chemical is in the sample (quantitative). The two kinds of analysis are not independent; a qualitative test will have a limit of chemical content below which it cannot detect the chemical. Therefore a positive qualitative statement is also a quantitative statement in that it says the compound is present at a concentration above the detection limit, which ought always to be stated. Conversely, a quantitative statement is meaningless without the qualitative identification of the chemical. It is very easy to fail to see that two related chemicals are distinct entities and, from this failure, to measure them as being one component. For example, many of the simpler and more useful analytical methods are sensitive to a class of compounds, they are not specific to one chemical. Thus the simpler methods of measuring trichloroethylene (a dry-cleaning fluid) will measure the concentration of many other chlorinated hydrocarbons that may be present.

Some of the above points can be illustrated by the history of dioxin as a contaminant of 2,4,5-T. Before 1957 the contamination of the herbicide 2,4,5-T with dioxin was not an issue because dioxin was unknown. When it was established that the ability of 2,4,5-T and related chemicals to cause the skin disease chloracne was a result of the contamination of those chemicals by dioxin, it was still not possible to estimate the quantity of dioxin in the 2,4,5-T. This was changed in 1966 with the development of a gas–liquid chromatography method to separate and measure dioxin. This method has been further refined, so that the detectable limits of dioxin have been progressively decreased as the sensitivity of the method has improved. The regulatory limit for dioxin content in 2,4,5-T has closely followed the downward trend in detectable limit; 10 ppm to 1 ppm to

0.01 ppm at present. The regulatory authorities have had to follow the analysts. You cannot set a permissible limit to the concentration of an impurity which is below the concentration you can detect. Therefore in many cases the permissible limit is easy to set; the analyst has done it for you.

Some legal authorities can ignore this principle. In the State of Victoria (Australia) there is legislation which reduces the permissible limit of alcohol in the blood of drivers to zero, that is, for learner drivers and those of less than 3 years driving experience. In technical terms this is nonsense. There will always be some alcohol in blood, 0.001% or 0.0001% or something, even if this is well below the presently permissible 0.05%. Whether this very small concentration is detected or not depends solely on the sensitivity of the analytical method. In practical terms there is no such thing as zero content. All one can say is that the compound is not present at a concentration greater than the detection limit of the particular method being used (which must be specified).

The reader will begin to see some interesting complications and interrelationships between toxic effect, dosage, sensitivity and reliability of analysis, etc. The question 'Is the product X contaminated by the poison Y, or isn't it?' becomes another of those simple questions that has no simple answer. This will further strengthen the public image of the scientist as a shifty, indefinite, evasive character who can't answer a straightforward question. Before we consider the implications of this situation further, it is worthwhile surveying the techniques of analysis to gain a further understanding of what analysis will tell us.

My own experiences over 30 years cover a broad enough timespan to illustrate the growth in analytical techniques. In the mid-1950s I measured the lead content of motor fuel by converting the lead compounds to lead chromate, combusting off all the hydrocarbon compounds, then weighing the lead chromate. Phosphorus in samples was measured by oxidising away all organic residues with perchloric acid, then measuring the colour intensity of a combination of the remaining phosphorus with molybdic acid, a complex of an intense blue. Such gravimetric and colorimetric methods were the basis of analysis. Chromatography was still a new technique, performed on packed columns or on paper. It is not in itself a quantitative analytical method, but is a separatory method which allows the components to be measured by techniques which will not work on crude mixtures. Thus we separated mixtures of amino acids by paper chromatography, then estimated the concentration of each component by reaction of the separated amino acids with the reagent ninhydrin. This produced a purple colour, the intensity of which was related to the quantity of amino acid present. The peak of efficiency in separation of components was perhaps the technique of ion-exchange chromatography as developed by Moore and Stein, who could separate a synthetic mixture of 50 amino acids and related compounds. At that time instrumental methods of analysis were infrared and ultraviolet spectroscopy. New methods were being pioneered: gas chromatography, mass spectrometry, electron spin resonance, nuclear magnetic resonance and thin layer chromatography were emerging.

By the 1960s the above techniques were established as general laboratory methods, and much effort was being expended to improve them. Now, in the mid 1980s, the sophistication of the instrumentation associated with these techniques has increased greatly, improving sensitivity, selectivity, accuracy and the ease of use of the instruments. New techniques have developed; for example, high performance liquid chromatography now complements gas chromatography as the two methods of choice for the analytical chemist, particularly in the detection and analysis of trace constituents as in environmental pollution control or in food quality investigations.

The ultimate method for trace analysis now is the combination of gas chromatography as a separatory technique with mass spectrometry as a means of detection and identification (known in brief as GC–MS). In order to give the reader an idea of what is involved in modern analysis, we will consider each step in the analysis of a crude sample by GC–MS. This is shown diagrammatically in Fig. 6.1. The analyst receives a crude sample and is told what class of compound to look for. It may be a post mortem sample of human tissue which is being examined for heroin, or a sample of soil suspected of being contaminated with dioxin. As a first step, the analyst has to extract the sample with a liquid solvent known to dissolve the compound being sought. This is done to bring the sample into a homogeneous form. After this, it may be possible to include a concentrative step into the procedure by evaporating off the solvent and then redissolving the residue in a smaller volume of solvent.

At this stage the extract may still be so messy that it cannot yet be put into the GC–MS. There are many clean-up procedures that can be used. One method is a form of liquid chromatography. The liquid extract is passed down a column of packed absorbent that absorbs the impurities and passes the chemicals of interest, or alternatively it absorbs the latter and allows the rubbish to pass through. In the latter case, the material for analysis can be desorbed from the column by a change of the solvent liquid. Further concentration is possible at the clean-up stage. If the chemicals of interest are reasonably volatile, they may now be injected into the GC–MS.

Relatively involatile components must be changed to volatile derivatives before injection. This is done by chemical modifications which mask those parts of the molecule associated with its lack of volatility. Such masking groups can be chosen so that they also confer a chemical label on the molecules of interest. For example, addition of a trifluoroacetyl group to a hydroxyl group of a sample molecule adds an easily recognisable group of three fluorine atoms to each sample molecule.

Now the analyst is able to inject the sample into the gas chromatograph. It is vapourised and carried down the separatory column in a stream of gas.

Fig. 6.1 Steps in the work up of a sample and its analysis by gas chromatography – mass spectrometry (GC–MS).

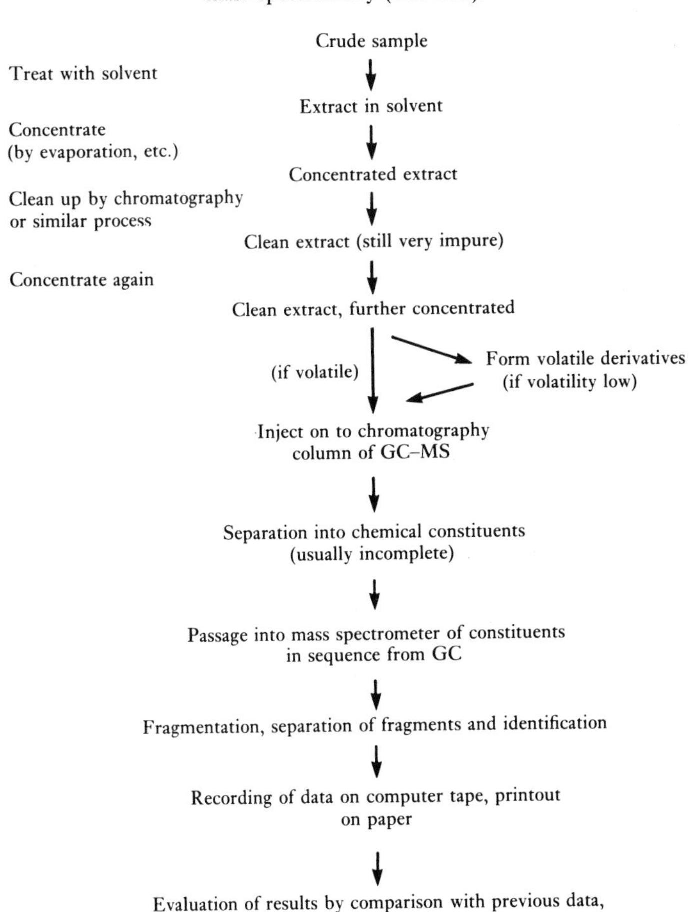

The components of the sample are separated as a result of differential absorption and desorption from a liquid film coating the column. The stream of gas and vapour is then directed into the mass spectrometer. At this point a great deal of separation has occurred. Modern chromatographs are much more efficient than those devised by Moore and Stein mentioned above. However, although the sample was already partially cleaned-up, the material injected into the chromatograph could still contain many hundreds of compounds, when the detection level is of the order of nanograms (10^{-9}, or thousand millionths of a gram). Although the chromatograph will have partially resolved these components, it is extremely unlikely that any one component will be emitted from the chromatograph in a pure state. Resolution is as a function of time, i.e. some components will come out more quickly than others. The analyst will know the retention time corresponding to the compounds being sought, because this will have been determined previously by chromatographing samples of the authentic compounds. Therefore attention can be concentrated on the mass spectrometer's response to the chromatograph effluent at or about this retention time.

In brief, the mass spectrometer can produce molecular, quasi-molecular or fragment ions in a predictable fashion. The fragments are then separated according to their charge to mass ratio, and the proportion of fragments of each mass can be detected and recorded. From the fragmentation pattern (which can be very accurately recorded) the chemical structure of the sample molecules can be determined. This is a qualitative procedure, but may be made quantitative by the inclusion in the sample of standard quantities of chemicals similar but not identical in structure to those chemicals being investigated, or an isotopic species of the same chemical. This is known as 'spiking'. In favourable circumstances the analyst can therefore confirm the structure of the compound being sought by GC–MS, and also quantify its presence in the sample.

Further qualitative confirmation of a component can be performed by a second sequential mass spectrometric procedure. The fragments from the first analysis can be further broken down into 'daughter' ions and again separated, allowing much better confirmation of the identity of the parent fragments (this double procedure is known as MS–MS).

At the end of the instrumental line is the data-processing equipment. The primary information from the instrument can be converted by computer into any form desired and can be printed as digital or graphical information. It may be stored on disc for future reference, or subjected to detailed comparison by computer with information called out from a library of previously acquired data.

The sensitivity of the GC–MS varies according to the nature of the sample and the skill of the operator. The instrument can be expected to detect and quantify 0.1 to 1 ng (1×10^{-10} to 1×10^{-9} g) in one sample injected. Given an original sample weight of 1 g and accepting (for practical reasons, including loss during workup) that perhaps one tenth of the sample extract can be injected, then the sensitivity of the assay is 0.1–1 ng in 0.1 g, or 0.001–0.01 ppm of the original crude sample. If concentrative steps can be included in the clean-up, and if larger original samples are available, then the sensitivity can be increased further. It is no coincidence that these concentration levels are similar to the regulatory levels for permissible dioxin content quoted at the beginning of this chapter.

This account of GC–MS is intended merely as an illustration of one of many modern instrumental techniques. There are many others using various chemical and physical principles to separate, identify and estimate chemicals in crude mixtures, which are well described by Ewing (1). It must clearly be recognised, however, that one cannot assume that because most of the analytical work occurs within the closely controlled instrumentation, therefore one can abandon the basic principles of the analytical chemist. To some extent, analysis is an art but the requirements are firstly a rigorous adherence to the standardised method of analysis, and secondly the careful exclusion of all factors that might interfere with the assay procedure. The checking of a procedure is best done by comparison between laboratories. A central organisation will prepare suitable samples of known composition and send identical specimens to each laboratory, with a code number as identification. Each laboratory will analyse the samples by the standard methods and report back to the central organisation. This body will collate all the results and compare them with the known composition, and with the mean values from the laboratory reports. Variation between laboratories is to be expected, and the allowable variation can be determined. Any estimate outside the allowable limits must be questioned, and the laboratory responsible will have to re-examine its procedures. Such interlaboratory checking is most important when an estimate of chemical composition is a requirement in a legal case, or for the enforcement of government regulations.

The sensitivity of modern analytical equipment demands greater care from the analyst, rather than less. I remember being given a partial chemical structure by our mass spectrometrist that did not look at all like what I had expected in the sample I had previously given him. I was further mystified when I found that the particular compound he believed to be present had not been mentioned in *Chemical Abstracts* in the last 10 years, i.e. it was a compound that had not attracted any commercial or research interest. Later the analyst came back to tell me that the compound was not in my sample at all, but that sufficient vapour had desorbed from his hands into the mass spectrometer when he was introducing my sample to give a signal on the instrument. The compound had actually been present in a previous sample from a completely different project, and the contamination had been carried over on the hands of the analyst.

Similar problems arise from the contamination of glassware used for storing or processing samples. Glass does not have an inert surface (the chemistry and physics of the glass surface are quite complex) and as a result, the surface will bind many types of chemicals, particularly positively charged ions and neutral molecules which combine a degree of polarity with lipophilicity (partly water soluble, partly fat soluble). Trichothecene mycotoxins are examples of the latter, and contaminate glassware so effectively that it is better to use the glassware once and throw it away, rather than to try washing it. This property of the trichothecenes may explain some of the observations in the Yellow Rain controversy (Chapter 10). We have had problems with dilute enzyme solutions in glass tubes, in that the enzyme (a large protein molecule) seemed to adsorb onto the glass wall. However, when we changed to disposable plastic tubes, other strange effects occurred. These seemed to be due to the release of very small amounts of a mould-release agent (a compound used to facilitate the removal of the newly formed tube from the mould), which partially inhibited the activity of the enzyme. In this case, the enzyme was acetylcholinesterase and the activity is the hydrolysis of acetylcholine, an event of importance in the transmission of nerve impulses across synapses. We went back to the use of glass tubes – you can't win them all.

The analyst therefore has to guard against two events. Firstly, the appearance in the sample of chemicals which were not originally there, due to cross-contamination from glassware, solvent residues or any other material contacted by the sample during workup and analysis. Secondly, all or part of the chemical which is under examination in the sample may disappear onto the surface of the glassware or whatever, and the analyst then reports that none of the chemical is present or gives an estimate that is far too low. This situation must be controlled by appropriate blank samples, i.e. the ability to find nothing in samples containing nothing (or more strictly, chemicals below the detection limit), and by known standard solutions, i.e. the ability to find in the sample what you have previously put there.

One important advance in instrumentation is the development of computerised data-processing equipment to go on the output end of the instrument, as mentioned briefly above in relation to the GC–MS. As well as being able to store, sort and compare data with libraries of information, such equipment can increase the sensitivity of the assay. The main problem to be solved if one wishes to increase sensitivity is to optimise the signal-to-noise ratio, that is to pick out and enhance useful information from the random, meaningless fluctuations that are derived from a wide variety of causes not related to the sample you have introduced. The signal-to-noise ratio can be enhanced by a number of methods, which are facilitated by computer equipment. Because this equipment is now cheap and very sophisticated, the output from analytical instruments can be subjected to quite complex processing. A general review of the use of computers in analysis and other areas of chemistry is given by Lykos (2).

A growing requirement for analysis is to monitor the concentration of chemicals in the workplace. To be realistic, this must measure the accumulated dose to a worker during the work day. It is no use doing occasional measurements; by chance you might strike times when the hazard is low, but a minute later the chemical concentration shoots up due to a machining operation, or to a spraying process commencing. Similarly, it is no use doing a measurement at a place in the workshop remote from the worker. Identity of the sampler with the worker both in time and place is essential. This requirement has led to the development of monitors small enough to be carried on the worker, with the collecting part near the worker's head. Sampling by such monitors is started when the work shift commences and finishes as the worker finishes. They are of two types. The active type has a small, battery-powered airpump which draws air into a trapping tube (commonly charcoal or other absorbent). The air flow is kept at a steady, slow rate. The contaminant is trapped in the sample tube and analysed later. Passive monitors dispense with the pump, and hence can be very small, like a badge. They consist of an absorbent disc behind a diffusion screen. The function of the latter is to reduce the transient effects of air currents and other disturbing factors, and give a closer relationship between the concentration in the air and the quantity absorbed by the disc in a given time. The sample collected by either type of monitor is assayed by gas chromatography or a similar technique after desorption from the trapping medium by the use of a suitable solvent. More elaborate monitors (Fig. 6.2) can give a direct reading of vapour concentration without a time delay for analysis.

These devices allow the measurement of the average concentration of a hazard in the workplace over a working day, and can therefore give data directly relevant to the regulatory levels, commonly given as Threshold Limit Value (TLVs). The TLV is the time-weighted average concentration over an 8 hour day below which concentration the worker is considered to be safe. The use of the TLV presupposes that the 8 hour shift is followed by 16 or more hours away from the working environment, and further that the effect of the chemical is gradual and cumulative. If the chemical can have a short term, acute effect, then a ceiling concentration ('*C*' limit) must also be stated. Some representative TLV and *C* limit values are given in Table 6.1.

Is this long exposition on analysis of any value to the average person who merely wants to be able to understand discussion and arguments on pollution, contamination of foodstuffs, industrial hazards and so forth? I believe it is. An understanding of how analytical figures are determined will destroy any blind acceptance of numbers as numbers, and help us understand what they really mean. A number presented with an estimate of variation is a shorthand way of expressing a complex situation. One must realise that in obtaining it the truth is inevitably compressed, sometimes unjustifiably so. In the latter case, the distortion of truth may be

Fig. 6.2 A monitor is being used to detect and measure potentially hazardous concentration levels of solvent vapours. This equipment (the Century 'Organic Vapor Analyzer') permits the direct readout of total organic vapour concentrations on the dial held in the operator's hand. Note the sampling probe, which is to be seen below his hand. Greater accuracy, and the separation of sample components can be done with the addition of further equipment (photograph courtesy The Foxboro Company, Foxboro, MA, USA).

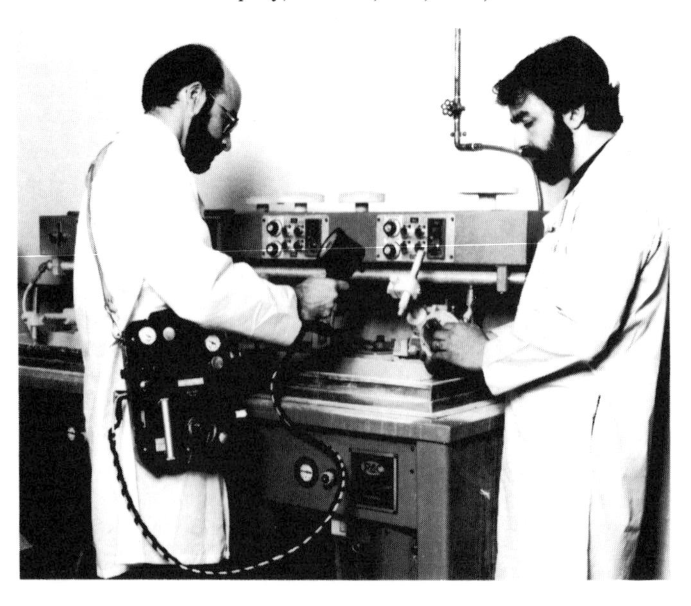

inadvertent, or it may be deliberate. Think carefully about what the numbers really mean.

It is difficult to conceive clearly what the numbers do mean when trace concentrations are involved. My favourite method of expressing and visualising the various concentration levels is in terms of length, which I believe is the easiest analogy to grasp. Close your index finger onto the thumb, then open very slightly. The smallest gap you can maintain is about 1 mm. A part per million is this distance in 1 km; a part per billion (ppb, i.e. a US billion) is the same distance in a 1000 km, or the distance a jet-liner flies in an hour. Or visualise 1 ppb as an ounce in 27 900 tons, if you can (this is added for our imperial and US friends, who cling to archaic complexity). When your analyst next quotes you a concentration of 1 ppm, also ask what the other 999 999 parts are. After the rude answer, ask what the major components are. These can have quite an important bearing on your one part of immediate interest.

Reference 3 illustrates the problems attendant on the question 'Is it there or isn't it?'. It discusses the problems of getting consistent results on one sample when the chemical in question is only present at the parts per

Table 6.1. *Some representative values for threshold limit values (TLV) also known as maximum acceptable concentrations (MAC) or hygienic standards*[a]

Acetic acid	10
Acetone	1000
Ammonia	25
n-Butanol	50 (ceiling)
Chlorine	1
Chloroform	25
Ethanol (common alcohol)	1000
Formaldehyde	2 (ceiling)
Hydrogen sulphide	10
Methylisocyanate	0.02 (skin absorption also possible)
Osmium tetroxide (as osmium)	0.0002
Phosgene	0.1
Pyridine	5
TDI (toluene diisocyanate)	0.02
Toluene	100
Vinyl chloride	10

[a] These values are those concentrations in air regarded as safe when averaged over an 8 hour working day. Ceiling values must not be exceeded, even momentarily. Concentrations in parts per million (ppm). **Warning**: do not use these figures for regulatory purposes. They may well be out of date, or not acceptable in your situation.

billion level. One laboratory may say it is there, and another may say it is not. This is a reflection of the great variation between laboratories to be expected at this concentration level. Figure 6.3 is taken from this paper by Horwitz and coworkers (3). It shows the increase in the variation of results between laboratories which occurs as the concentration level of the chemical falls. This is for laboratories experienced in the relevant analyses, not for first-time, one-off analyses. The coefficient of variation or percentage standard deviation (see definition later) at the parts per billion level (in an assay of aflatoxins for example) is about 50%. Therefore a mean value of 2.6 ppb would result from a number of determinations, two out of three of which would be in range of 1.3 to 3.9 ppb. One out of every three results would be expected to be further from the mean.

The analyst recognises the fact that his quantitative answers are uncertain by estimating the degree of uncertainty and including that in his answer. This is commonly done by taking the mean of a number of determinations and calculating from each individual result an expression

Fig. 6.3 The increase of the interlaboratory coefficient of variation as a function of reducing concentration. The data used here were obtained from interlaboratory collaborative studies made by the Association of Official Analytical Chemists over the past 100 years. For further details see the original paper by Horwitz (3). (Reprinted from (3) with the permission of the Association of Official Analytical Chemists, Arlington, VA 22209, USA).

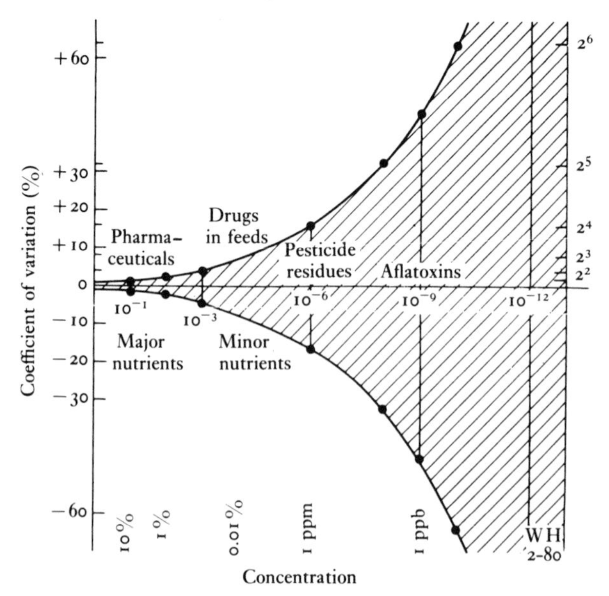

of variation or deviation. A popular mode of expression is the standard deviation (see reference 4), so that the concentration of the chemical X in a mixture may be reported as 142 ppm $\pm$ 21 SD (11). This means that the analyst made 11 (the number in brackets) separate determinations on the same sample, that the mean of these 11 results is 142, and from the scatter of the results about the mean he estimates that the standard deviation (SD) is 21. What does this mean? Simply, it means that 68% of the determinations lie within one standard deviation of the mean (i.e. they lie between 121 and 163 ppm). This is not very certain; there is one chance in three that a determination lies outside this range. A range of two standard deviations includes 95.5% of results, that is between 100 and 184 ppm in the above example. The SD simply describes the variation the analyst has found in the results. It gives some indication of the reliability of the results, that is, in the ability to get a consistent result, which is 'precision' in the analyst's terminology. It says nothing about 'accuracy', which is the ability to get near the true value, as determined by other independent evidence. The concepts of precision and accuracy are best illustrated by a target diagram, as in Fig. 6.4. On the left the five shots (or determinations) are close together but not near the bull's eye (the true value), thus achieving good precision, but with a directed error. In the middle diagram the shots are scattered widely, but they centre about the true value (accurate but imprecise). On the right is the ideal, in which the shots are close together and centre on the bull's eye. Precision is fairly easy to express as a standard deviation. The problem is with accuracy, which has to be established by means such as the interlaboratory checks of common samples which were described previously.

The analyst's figures are usually put in the public domain stripped of all conditional and limiting trimmings. The results will be reported by the

Fig. 6.4 The analysts take up archery. Mr A is quite precise but lacks accuracy; he must be consistently misaligning his arrows. Mr B is accurate but cannot maintain precision. Ms C has both accuracy and precision

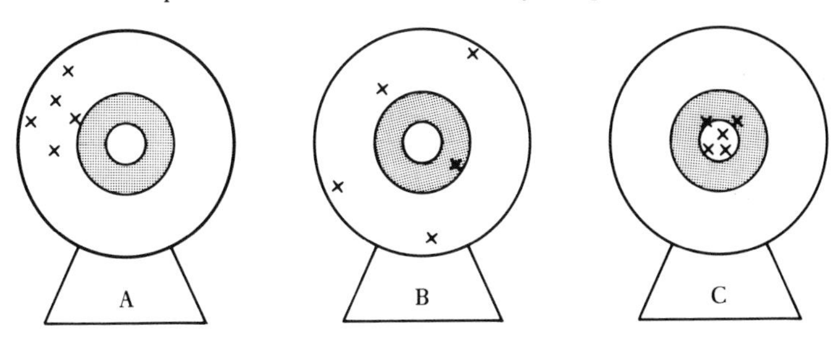

media as finding 142 ppm of *X* in the mixture. He may not have obtained 142 ppm in any of his 11 determinations, and the public has no indication of how precise or accurate the figure is. The value is more reasonably given as 140 ppm, for the final 2 has virtually no meaning given the variation in the individual results.

You can therefore ask for and expect to get some indication of the precision of a result. This will give some reassurance as to the meaning of an analysis. What about accuracy? Unfortunately, there may be no information at all. In the example above, the chemist tells you that there are 140 ppm of *X* in the sample, and can give you an expression of precision (± 21 S.D.). You have to form your own estimate of the likely accuracy of the result. If the chemist can show that the assay has been done many times before, if the laboratory is certified to do that type of assay by some official or professional body, and if he can quote participation in an inter-laboratory test, then the accuracy is likely to be good. If it is the first time he has done this assay, then beware! It could mean anything.

When you consider precision in a real-life context, you have to decide how relevant it is to the matter in hand. In the above example, if the regulatory limit for *X* in the sample were 15 ppm, then the precision about the found value of 142 ppm is hardly relevant; the sample will obviously fail the regulatory requirement. To a degree, accuracy is not relevant in this example either, although inaccuracies by a factor of ten are not uncommon. Other situations may demand high precision and accuracy, it depends on the purpose of the assay.

From the foregoing, you may now have got the idea that all analyses are meaningless. This is not so. Most are well-conducted affairs performed by qualified persons who are constantly aware that their fellow chemists are peeping over their shoulders and checking what they are doing. I hope that the above discussion has given you the ability to spot the occasional rogue.

Uncertainty is, however, implicit in analysis, especially at low levels (Fig. 6.3). Therefore there is a forest of uncertainty between certain high concentrations and zero concentration. This forest is populated by chemists seeking to improve techniques, lawyers looking for law suits, government officials seeking to set regulatory standards, the media hunting for sensations and by the general public who are more or less completely lost. This chapter aims to give you some guidance through this forest.

The partial answer to the question 'Is this chemical toxic or not?' is that the toxicity depends on the dose of the chemical (Chapter 4). Now we find, however, that the dose of the chemical may not be certain. This applies particularly if the toxic chemical occurs as a small fraction of a mixture with others of lesser toxicity. The toxicity of 2,4,5-T depends on the content of dioxin (Chapter 13). If the dioxin content is variable and difficult to measure, we cannot make very accurate predictions about the toxicity of a sample of 2,4,5-T. I am sorry to make matters appear even more uncertain and complex than was apparent at the end of Chapter 4. However, that is the nature of the real world. There is no completely non-toxic chemical; there is no totally pure chemical. There are no absolutes, for we live in a relative world.

References

1. Ewing, G. W. (1975). *Instrumental Methods of Chemical Analysis.* Tokyo: McGraw-Hill Kogakusha.
2. Lykos, P. (1982). Computers in the world of chemistry. In *Advances in Computers*, vol. 21, ed. M. C. Yovits, pp. 275–331. New York: Academic Press.
3. Horwitz, W., Kamps, L. R. & Boyer, K. W. (1980). Quality assurance in the analysis of foods for trace constituents. *Journal of the Association of Official Analytical Chemists*, **63**, 1344–54.
4. Rowntree, D. (1981). *Statistics without Tears. A Primer for Non-mathematicians.* Harmondsworth: Penguin.

7

Chemical hazards: the background to real and imaginary risks

We cannot evaluate those hazards to society which are due to chemicals until we have surveyed those which have always been there, and are therefore not due to the New Chemical Age. Therefore I want first to survey this background of pre-existing risks, then see what may have been superimposed on it by the modern use of chemicals.

When we attempt to establish what people do suffer most risk of being injured or of dying from, we find the problem that the most reliable statistics are the most recent ones, and these are possibly influenced by the Chemical Age. Nevertheless, we can disentangle those hazards which obviously have nothing to do with chemicals. There is a wide variety of statistical data to choose from, which are not all comparable and need careful interpretation (mainly concerning what population and sub-population they are drawn from). Table 7.1 is from a US population (1). The figures in the first column are expressed in a way which weights the figures in a socially meaningful way, whereas the annual rate figures in the third column are not so revealing.

Accidents and adverse effects caused the greatest loss of potentially useful life, although they were fourth in terms of mortality rate. This is because accidents are more common among younger persons compared, for instance, to diseases of the heart (which were first in rate of mortality but third when weighted and are complaints of older persons). With reference to Chapter 8, note the prominent position of cancers (malignant neoplasms) and the relatively insignificant position of infectious disease, represented by pneumonia and influenza. The not very startling conclusion from Table 7.1 is that young people are more likely to die from car accidents, suicides and homicides, whereas the older people die of diseases of the heart and circulation. Of course some young people die of heart disease, strokes, leukaemia or pneumonia, but the proportion of deaths by more violent means is higher than in older groups of the population. Similarly some old people will die in car smashes. Other western societies show similar patterns, with less emphasis on homicides than in the USA. This survey refers to deaths (mortality); we are also interested in sickness that does not necessarily result in death (morbidity).

These statistics are often harder to collect because, although many countries now have reasonably reliable registrations of death with cause, records of disease incidence are less well maintained and more difficult to compile. We often therefore have to rely on mortality data and hope they also reflect morbidity. This can be a dangerous assumption and must be explored before it is accepted. Consider a simple example. Most deaths in car accidents in the State of Victoria occur in the country. The initial conclusion is that the country roads have the greatest traffic accident problem. This is not so. The biggest problem in terms of personal injury and car destruction is in the towns, principally Melbourne. This is because the proportion of deaths per accident in the country is much higher than in the town. The country speed limit is 100 km/h, which means many cars travel at 110 or 120 km/h. If they slam into a tree or an oncoming semitrailer, death for the driver is almost inevitable. In the city, most crashes occur at intersections or are nose-to-tail impacts at relatively low speed. These produce injuries which can be quite serious and are much

Table 7.1. *Years of potential life lost, deaths and death rates by cause of death for the United States*

Cause of death	Years of potential life lost before age 65 by persons dying in 1982	Estimated mortality September 1983	
		Number	Annual rate/100 000
All causes (total)	9 429 000	157 500	818.6
Accidents and adverse effects	2 367 000	8 120	42.2
Malignant neoplasms	1 809 000	37 020	192.3
Diseases of heart	1 566 000	56 500	293.5
Suicides, homicides	1 314 000	4 230	22.0
Cerebrovascular diseases	256 000	12 000	63.4
Chronic liver disease and cirrhosis	252 000	2 140	11.1
Pneumonia and influenza	118 000	3 210	16.7
Chronic diseases of lungs	114 000	4 410	22.9
Diabetes mellitus	106 000	2 950	15.3

See (1).

greater in number than deaths on country roads. Many forms of disease can also be socially very significant although they produce few deaths. Mortality is not therefore a good indicator of morbidity, but often it is all we have.

We are also concerned about events before birth, since deformities at birth are an alarming occurrence. The egg is fertilised by sperm, then begins to divide and the embryo may successfully attach to the wall of the uterus, or it may fail to do so and be lost. The first third (trimester) of the embryo's life in the uterus is its most sensitive period, as the main organs of the body are being formed. Quite often something goes wrong; the embryo just fails to develop beyond an early stage, or it becomes grossly deformed. This may be due to internal influences, or to external ones (including chemicals). The body's answer to this is to reject the embryo, and spontaneous abortion results, which is not too dramatic an event when it occurs early. The proportion of spontaneous abortions is hard to estimate; perhaps 20% of conceptions are aborted early (15% of all identified pregnancies is one estimate, reference 2). Sometimes a deformed embryo is not rejected, and is carried to full term and born as a deformed child. This may be regarded as a failure of the body's normal rejection procedure; for some reason the embryo was not recognised in the uterus as being abnormal. The deformities at birth vary enormously in degree. Probably none of us is 'normal' for we all have minor defects. The proportion of gross abnormalities at birth is around 1 – 2% of births (Table 7.2), depending exactly on where you draw the line between spina bifida, Down's syndrome and severe cleft palate on one side, and missing single digits, minor mental retardation or harelip on the other.

It follows from the above that any population will show between 1% and 2% of children with defects (gross or relatively minor) at birth. If you are surveying a population of, say, 10 000 live births to see if there is a greater than usual number of malformations, then you will find there are around 100 to 200 children who do show such defects. You then have to find definite indications of an excess number over and above this background value. This may be impossible unless your sample population is very large, as you are the victim of chance, random variation in the numbers actually found. Thalidomide was only recognised as a cause of birth defects because those defects caused by it were very unusual. Had the defect caused by thalidomide been a common one, then the thalidomide effect would have taken much longer to discover. This topic is further considered in Chapters 13 and 14.

We can now revert to adults, and try to ascertain the proportion of mortality and morbidity that can be ascribed to environmental and occupational factors, particularly to chemicals. Looking at Table 7.1, we can cut out many types of accident, suicides, homicides, pneumonia and influenza as not being anything to do with chemicals, although a case can be made for considering chemical exposure to be a possible indirect cause of homicide or, more probably, suicide. Next we put aside cancers, about which we can have terrific arguments but I have considered this question separately in Chapter 8. We are then left with the problem of trying to determine the contribution of chemicals to the causation of some accidents, heart, cerebrovascular, liver and pulmonary diseases and diabetes. We can do this by surveying populations particularly likely to be exposed to chemicals and comparing them with other populations in which exposure is negligible. This is possible when the source of exposure to chemicals is occupational, i.e. at the place of work. It is harder when the possible exposure is domestic or environmental.

Pochin (4) has surveyed fatality rates relating to occupational factors and his paper is well worth reading as background to the topic. Table 7.3, taken from his paper, is specifically concerned with exposure to chemicals. This is largely a historical survey not applicable to modern conditions; the examples given are of rather obvious diseases affecting small, specialised groups of workers. Thus the high rate of deaths in 2-naphthylamine

Table 7.2. *Reported incidence of congenital malformations in the United States in 1980*

Malformation		Rate per 10 000 total births
Spina bifida and anencephaly		8.5
Heart malformations		30.0
Hydrocephalus (without spina bifida)		4.3
Cleft palate and cleft lip		12.9
Clubfoot		25.5
Reduction deformity		3.8
Defects of intestines and/or trachea		5.4
Kidney inadequacy		1.2
Down's syndrome		7.5
Hypospadias (per 10 000 male births)[a]		51.1
Rates for all births (excluding hypospadias)	=	99 (0.99%)
Rates for all male births	=	150.1 (1.5%)

Modified from (3).

[a] Hypospadias is a condition in which the opening of the urinary tract is displaced from the top of the penis, and the genital organs are in a varying degree of underdevelopment.

manufacture (24 000 per million workers per year : 2.4% per year) is now historical and refers to a small actual number of workers (see also Chapter 8). What we are now considering is the science (or art) of occupational epidemiology. This is a topic fraught with difficulty, extremely demanding of care in interpretation, and consequently the source of endless argument. Enter it cautiously. As a basic introduction, Monson's book (5) is excellent in establishing the methodology.

Table 7.3. *Estimated rates of fatality of disease attributed to types of occupational exposure*

Occupation	Cause of fatality	Rate (d/M/y)[a]	Standard error
Shoe industry	Nasal cancer	130	35
Printing trade workers	Cancer of the lung and bronchus	About 200	40
Workers with cutting oils			
Birmingham	Cancer of the scrotum	60	—
Arve district	Cancer of the scrotum	400	50
Wood machinists	Nasal cancer	700	200
Uranium mining	Cancer of the lung	1 500	
Coal carbonisers	Bronchitis and cancer of the bronchus	2 800	650
Viscose spinners (ages 45 to 64)	Coronary heart disease (excess)	3 000	1000
Asbestos workers			
Males, smokers	Cancer of the lung	2 300	750
Females, smokers	Cancer of the lung	4 100	1150
Rubber mill workers	Cancer of the bladder	6 500	3400
Mustard gas manufacturing (Japan 1929–45)	Cancer of the bronchus	10 400	2200
Cadmium workers	Cancer of the prostate (incidence values)	14 000	8000
Amosite asbestos factory	Asbestosis	5 300	1400
	Cancer of the lung/pleura	9 200	1800
Nickel workers (employed before 1925)	Cancer of the nasal sinus	6 600	1050
	Cancer of the lung	15 500	1500
β-naphthylamine manufacturing	Cancer of the bladder	24 000	2700

Adapted from (4) which should be consulted for sources of the information and a discussion of its meaning.

[a] d/M/y = Deaths per million exposed persons per year.

You may think I am dodging the question of how much disease is due to chemical causes. I am. I will discuss the matter in various parts of this book, and I believe that overall you will gain an insight into the difficulties of such an estimation. Here I want to outline the methodological problems. In fact, I cannot treat epidemiology adequately here, and refer you to Monson (5) for detail. As an illustration of the problems of interpretation, we can consider factors which confound epidemiological surveys because of their gratuitous relation to the factors which are important. Consider the example of heart disease in Britain (6). The incidence is higher in Scotland and northwestern England than in southwestern England. Why? We don't know. It has been suggested that the higher incidence of the disease is caused by a variety of factors, as listed in Table 7.4. Several factors probably combine to produce the effect. What makes epidemiological analysis difficult is that these factors have relationships among themselves. Suppose (purely for example) drinking habits were the cause of the problem. This may relate to the rate of unemployment in the North and West (which is high), which in turn is related to depressed economic conditions. Thus it could be possible to show a statistical relationship between the number of persons per room in dwellings and the rate of heart disease. Such an association is not between cause and effect.

Consider another example, again from a *New Scientist* item (7). A study in Massachusetts found an association between deaths due to tumours of soft body tissues and service in Vietnam. The incidence of tumours was higher than in a group of servicemen who had not been in Vietnam. One of the researchers is reported as saying that there was not much difference between the two groups except that the Vietnam group was exposed to

Table 7.4. *Factors associated with living in Scotland and northwestern England*[a]

Weather
Drinking water
Air quality
Racial type (i.e. Celtic versus Anglo-Saxon)
Rate of unemployment
Other depressed economic conditions
Smoking habits
Drinking habits
Diet

[a] The comparison is between Scotland and northwestern England which have a high incidence of coronary heart disease and southeastern England, which has a lower incidence.

defoliants. On reflection, one may think of a good many factors by which the group could differ, in addition to exposure to defoliants (not a universal condition of service in Vietnam; see Chapters 13 and 14). Some that occur to me, which are chemical, are listed in Table 7.5. Obviously it is foolish to select arbitrarily one factor from this list; we have to consider every one of them unless there is good reason to reject it.

The problem of seeing any disease in a population (whether chemically induced or not) can be graphically represented, as in Fig. 7.1. The vertical axis represents the percentage of the population showing a condition, whether congenital abnormality around 1–2%, or minor psychiatric troubles between 10–30%, or any other disease pattern. The horizontal axis is an arbitrary time scale representing the period over which information is available. The figure shows a continuous line to represent control data from an unaffected population and the broken line is the result of our observations on the population under test. In the top example the two are indistinguishable, as the test sample line is about as often below the control as above it. Any difference is unobservable (it may be there, but we cannot see it). The bottom example shows an increasing divergence with time, and arouses our suspicion that some factor is increasing morbidity in the population. It is suspicion only; we would have to ensure that this is not just some chance freak of normal variation.

This problem of distinguishing a real change from normal variation in a population is exactly the same as that which faces the analyst in distinguishing a real signal from instrument noise when performing analysis at very low dilutions (Chapter 6). The analyst has to try and

Table 7.5. *Chemical factors which could be associated with service in Vietnam*[a]

Exposure to defoliants
Exposure to antimalarial drugs
Exposure to insecticides
Exposure to insect repellents
Exposure to other medications (?)
Excessive alcohol intake, as a result of increased stress
Use of illicit drugs, as a result of increased stress
Pollution of water supply peculiar to Vietnam
Contamination of food supply peculiar to Vietnam
Pollution of air peculiar to Vietnam

[a] The comparison is between two groups of servicemen, with no apparent differences save service in Vietnam. (Many non-chemical factors could also be the causative agents of any diseases peculiar to Vietnam veterans.)

achieve optimum signal-to-noise ratios, and has more chance of doing this than the epidemiologist. The latter cannot easily take larger samples, nor can he increase the gain in his system or smooth out his no-signal baseline. The problems of the analyst and the epidemiologist both centre, however, on the difficulties of getting real data from uncertain situations. Interpretation is all important.

The conclusion is that it is very difficult to establish cause and effect by epidemiological means. The statistics do not lie. However, the answer given by statistical methods is only as good as the information fed in, on the GIGO (guff in, guff out) principle. Almost any conclusions can be obtained by suitable misrepresentation of data. Be careful. Take nothing on face value. Poke and ferret around a topic until you are sure the statistics do have value.

We should now consider another aspect of the occurrence of relatively rare diseases in a community. The occurrence may be random, but

Fig. 7.1 A hypothetical examination of an exposed population for an increase in a medical condition which occurs in 1–2% of the unexposed population. Results for a 10 year period are available. In the upper set (*a*), the exposed population (broken line) is not distinguishable from the control population (continuous line). In (*b*), the exposed population diverges from the control population with time. This may be a real effect, or perhaps a freak variation in the exposed population which will subsequently return to normal.

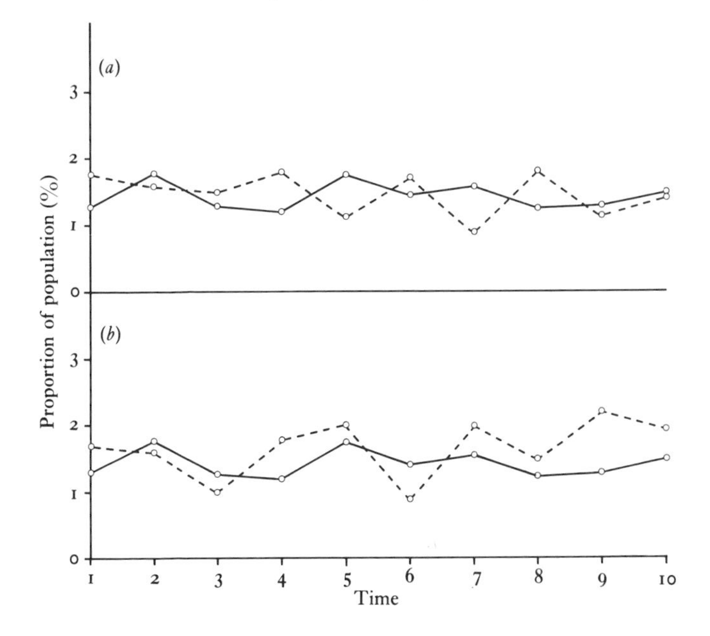

nevertheless several instances of the disease may appear in some way connected, in the same workplace perhaps, or in the same street. Does this mean that the disease is connected with the workplace, or the locality? Not necessarily: such clusters may occur in a random distribution, or they may be the result of a common cause in workplace or street. I will generate a random series of numbers by throwing a dice ten times. The result is: 5,1,6,5,5,5,6,4,1 and 6. There is a cluster of three fives. I throw another series: 6,5,6,2,3,1,1,1,2 and 3. Again a cluster of three identical digits (no, I am not cheating). A third and last set of throws gives: 1,2,1,2,4,3,3,1,3 and 4. A cluster of two threes and another three nearby. These are sets of random numbers, but quite obviously they have properties of order of their own. More exactly, they are randomly selected series; once selected they have identities and patterns. In fact it would be highly artificial if *no* clusters were found; this would imply an imposed order. Thus if a disease is randomly distributed in a community, there will be occasional clusters of that disease. If such a cluster occurs in a factory using a chemical process, we have the difficult job of deciding whether it is truly random, or perhaps a disease which is caused by one of the chemicals being used. We can only do this by examining the workers in other factories which use the same process. The original factory cannot be used in this confirmatory study, since we know we will find the disease there, and we are also aware that this has become a selected sample. Think about this a little; it is an interesting point. The topic is further discussed by Enterline (8) and Ennis (9).

The above discussion means that if a rare cancer kills two or three workers in a chemical factory, or a similar cluster of birth malformations is found in an agricultural community, there is no necessary cause and effect relationship between a chemical and cancer or abnormal children. Can such clusters be ignored? No, because there may be a connection. We have to examine the possibility of this by whatever means we can. At the same time we have to inform the community about the true situation in order to allay their natural fears, and to reduce the hysteria that the more sensational branches of the media will generate.

What do people feel are problems, as distinct from those we can shadowily see as really being troublesome? Upton (10) has presented interesting results from a survey of three groups of persons: members of the League of Women Voters; college students and business and professional club members. He lists 30 sources of risk (including smoking, nuclear power, commercial aviation) according to their actual annual contribution to the number of deaths in the US, then ranks them according to the perception of the three groups. If we select three chemically related factors, namely food colouring, food preservatives and pesticides, we find them at positions 26, 27 and 28 respectively of the factually determined list. As perceived by the three groups, the rankings are shown in Table 7.6. All three groups of people inflate the risk from pesticides. The students tend to rank all chemical factors higher than they deserve, the other two groups are quite realistic about food colouring and preservatives. The full data in Upton's article are worth reading and studying.

A second (11) quite different study is also informative. This was based on interviews with a representative sample of the adult population of England and Wales. The interviewees were asked a number of questions about their attitudes to a variety of industrial and other risks. The results are difficult to generalise or compress into simple statements, so I refer you to the original document for details. On the questions that related to chemical matters, it seems to me that the public have a fairly realistic idea of the risks. The estimate of the worst possible number of deaths from an industrial or chemical works disaster was spread very widely from less than 10 to 50 000 or more. Past actual experience suggests the lower figures are correct, but a study cited in the report (11) says a higher figure is possible. Remember, the interviewees were asked for worst case figures, and this was before the Bhopal disaster. Air pollution was a topic on which a large number of those surveyed said they had no idea how many deaths could be attributed to it. This was very wise, as this reflects the state of scientific knowledge. Nevertheless, those that did reply gave a low figure. The respondents tended to underestimate the number of deaths from lung cancer, but recognised smoking as a health problem. When asked what they were currently worried or concerned about, that is risks that could affect them or their families, the respondents listed traffic accidents, accidents in the home, and crime and violence as of particular concern. Chemically related risks were not prominent in their perception.

Table 7.6. *How three groups of persons perceive the risk from three chemically related sources of risk out of 30 sources*

Source of risk	Actual order of risk (out of 30)	Perceived order of risk[a]		
		Group A	Group B	Group C
Food colouring	26	26(0)	20(+6)	30(−4)
Food preservatives	27	25(+2)	12(+15)	28(−1)
Pesticides	28	9(+19)	4(+24)	15(+13)

[a] Group A, members of the League of Women Voters; Group B, college students; Group C, business and professional club members. Numbers in parentheses show change in ordinal places. Modified and summarised from (10).

My impression from these two surveys is that members of the public are being reasonably accurate in giving a low degree of relative risk to chemical factors, but that grafted on to this one may find particular concerns among intellectuals, which result from the concentration on one or two topics (e.g. smoking, pesticides) by the information media. What the media adopt as a cause has little to do with its impact on society in real terms, and much to do with its value as a sensation. It would be interesting to pursue this study further, i.e. what the public perceives as the degree of risk from the Chemical Age, as opposed to that promoted by the media.

The ability of the media to present relative risk to the public is the subject of a report (12) on the communication of scientific risk. The Task Force that produced this report believed that, given the constraints under which they work, the media have done a good job in reporting on technological risk. I believe they came to this conclusion as a result of consideration of a media system unknown to us. They considered an image of the media which the media people hope the public sees, not the media which we are in fact familiar with. Fortunately the chairman, Harrison Schmitt, produced a much more credible and penetrating dissenting report. Read the report for yourself, it is worth studying.

So far we have considered what we may term physical disease, but have neglected psychiatric disorders. We tend to dismiss such disorders of the mind as 'not real', but they are real and costly to society in terms of suffering, time lost to useful activity and time spent in providing care. The prevalence of such disease in a community is hard to determine, as we are looking for morbidity rather than mortality. Mortality figures for suicides and homicides (Table 7.1) give an indication of the desperate cases of psychiatric upset, but for the rest we are left to guess the prevalence from what little data are produced from various surveys. An estimate of 10% of the community as suffering from severe or marginal psychiatric disorder is probably not far from the truth. It depends on where you place the border between eccentricity, cynicism and depression on the normal side, and psychosis and neurosis on the abnormal.

It may be easier to define the problem in terms of what proportion of the population is kept from its full useful potential by psychiatric problems. By this definition, 10% may be too low. A study in the UK on younger civil servants (13) showed that 33% of them at any one time were suffering from minor psychiatric disorders, principally depression. A year later approximately half the previous cases had become symptom free, but another set of cases had taken their place. Thus there was not a permanent population of sufferers, most were passing through periods of depression which resolved themselves. At first I dismissed this report as being virtually without significance, as a proportion of 33% disturbed must intrude on normalcy; how could such a proportion be abnormal? Then I made the analogy with other disease, and appreciated what the figures meant. At any one time a fair proportion of us will be suffering from colds, minor aches or pains, cut fingers, blistered feet, all of which may affect our work and leisure activities to a small degree, but not be worth treating by a doctor. In the same way, minor psychiatric complaints have a small but significant effect on us; in the study cited (13) only 2–3% of the subjects were considered to need treatment. A healthy population thus includes a substantial proportion of people with minor physical and psychiatric complaints. We must take account of this when we wish to look for new disease patterns in a community.

Because of the widespread occurrence of psychiatric upset in the community it is difficult to detect factors which may increase it, particularly when the community in general does not accept that it exists. Whether chemical factors are involved, I do not know. Occupational exposure to chemicals which cause mental disorders (e.g. mercury salts causing hatters to go mad) is nowadays fairly well controlled, and most modern problems I suspect are due to the abuse of drugs, whether therapeutic or illicit. The point to remember is that any population contains unfortunate people with mental upsets. Therefore if anyone says that 10% of a certain group of persons who work with chemical *X* have mental disturbances and therefore *X* causes mental problems, ask for a detailed comparison with a large group of unexposed persons. Then find a competent statistician who can estimate whether the exposed and control groups are different or not. Our society wants to overlook congenital abnormalities and mental disease. We can not, as we want a true picture of our world.

In some degree related to the above is the possible occurrence of what is termed psychogenic illness, that is, an illness caused by psychological factors. The illness may well resemble one caused by some physical or chemical factor and thus be difficult to recognise. The most likely cause of such an illness is stress; the sufferer need not have a mental illness but simply finds him or herself under unusual pressures. Psychogenic illness is one aspect of the relationship between mind and body which is referred to as psychosomatism, or the influence of the mind (psyche) on the body (soma). I stress that psychosomatic problems are real; it is just that the cause is more remote from the situation we see than we sometimes perceive.

Psychogenic illness can be communicated among groups of persons to become epidemic. This may then be called, somewhat unkindly, 'mass hysteria'. The implied association with females is unjust (I am referring to the origin of the term hysteria) since, as pointed out in reference 14, those industrial episodes of epidemic psychogenic illness that have been reported occur among the poorly educated, most stressed part of the workforce, which is predominantly female. Sex may be a confounding factor in the relationship between this illness and its causes. This reference (14) describes episodes of nausea, headache and disorientation among workers at a plant manufacturing electronic components. Several investigations failed to find any chemical cause for these complaints (that is, no concentration of chemical vapours, lead, sewer gas or air-conditioning refrigerant) sufficient to be a problem. The sickness was ascribed to psychogenic illness. This diagnosis can never be completely certain but, in this case, it was strengthened by the fact that full information of, and communication with, the employees resulted in the abatement of the illness. Psychogenic illness is likely to arise where workers are socially stressed, undereducated and not in communication with management. They are aware of possible problems with chemicals, and therefore their stresses may find expression as their concept of a chemically induced disease. The remedy is with management; not to ridicule and suppress the problem, but to consider it fully and to talk freely with the employees about their sickness and its possible cause.

Thus in any consideration of a disease which has been ascribed on first sight to chemical exposure, we must consider the possibility of psychogenic illness. If we deduce the problem is indeed psychogenic, then we must ask for the appropriate solutions as outlined above. We must not dismiss it as 'all in the mind' and therefore not worth consideration.

We tend to forget that the hazards from chemicals are not all due to the toxic effects of chemicals. Chemicals can be flammable and explosive (Fig. 7.2). Fire in chemical factories and stores is a frequent occurrence which can be costly in stock loss and tragic when people are killed. It is also a problem in chemical laboratories. I was in the last year of students that used the old chemical laboratories in Cambridge, built in the 1890s. We used old bored corks to make up distillation equipment, and they leaked vapour. If this was flammable, fires were common. Also the benches were served by open lead drains running the length of the laboratory. It was not uncommon to see a yellow flame shoot along these drains as ether vapour caught fire. Explosions are even more dramatic, as evidenced by the explosion which destroyed a whole factory at Flixborough, and by a casual reading of industrial accident reports compiled by the various factory inspectorates and national safety bodies. Hay (15) in his discussion on dioxin problems makes the point that the most obvious injuries related to this substance are the mechanical ones suffered by factory workers when various trichlorophenol or 2,4,5-T plants have blown up. A brick on your head is an indisputable fact; chronic toxicity is hard to demonstrate. Similarly, in the laboratory, explosions are probably one of the main causes of injury. They occur most often during distillation of a liquid, when most of the pure liquid has been distilled off and what remains is a complex mixture. This occasionally is chemically unstable and detonates, showering broken glass over a wide area.

I have never seen figures which compare fire and explosion injuries in factories with corresponding effects due to chemical toxicity. Such information would be interesting, and I am inclined to bet the former outnumber the latter. In this respect we can contrast the two appalling disasters that occurred in late 1984; hundreds of Mexicans (officially 500) killed by explosions and fireballs caused by liquefied petroleum gas escaping at Ixhuatepec, and 2500 Indians poisoned by methyl isocyanate at Bhopal (see Chapter 18).

One aspect of the above hazards from chemicals is often neglected but, to me, worrying. Chemicals are often transported by road or rail and are therefore subject to accidental spillage or rupture of the container. There is

Fig. 7.2 Not all chemical hazards are due to toxicity. The fireball shown here is the result of the sudden release and ignition of liquefied petroleum gas (LPG) from a 15 kg cylinder. Such cylinders are common now as many vehicles have changed from gasoline to LPG as fuel. This photograph was taken during an experiment to measure the hazards from accidental explosions (from the 1981 issue of *Health and Safety Research*, produced by the Health and Safety Executive, UK).

Fig. 7.3 Methods of dispersing clouds of heavier-than-air gas are being investigated in Britain. They have relevance to the control of disasters such as that at Bhopal in 1984. Here a water-spray barrier is being used to disperse a gas cloud. The spray nozzles are 3 m above the ground; the wind direction is from left to right. (*a*) Spread of gas from source on left of barrier; (*b*) the scene shortly after the water spray was turned on; (*c*) the situation a little later: the spray has, in fact, acted as a barrier to prevent further movement of the cloud (from the 1981 issue of *Health and Safety Research*, produced by the Health and Safety Executive, UK).

(*a*)

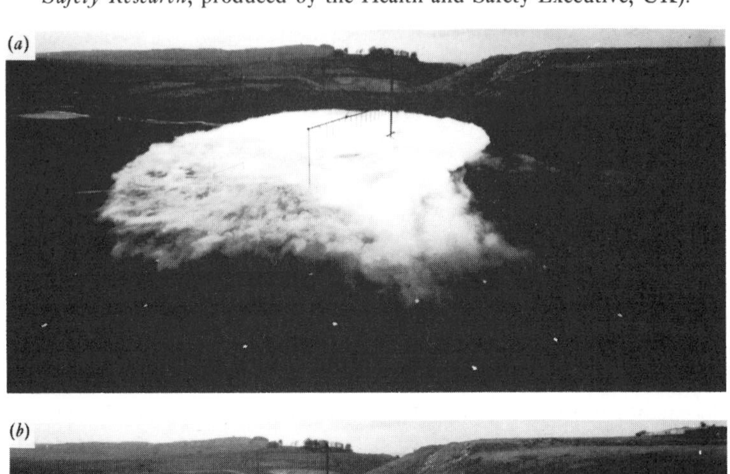

(*b*)

(*c*)

then a risk of fire, explosion or toxic contamination, or even of all three simultaneously. Because accidents to tankers carrying petroleum products are not unusual, we have become accustomed to them. The fire brigade comes to hose down the road, the police isolate the area for a time, and the reporters have a minor news item. Certainly the various fire brigade authorities are worried about the possible complex effects that could arise from more exotic chemical loads. Bizarre accidents are conceivable; having seen liquid nitrogen tankers on the roads quite commonly, I wonder whether any unfortunate pedestrian is going to be turned into an instant iceblock when a tank ruptures.

It may be worthwhile to go back again to consider the risks that have a major influence on people's lives, and therefore ought to be of primary concern to society. Cohen & Lee (16) have assembled a list expressing all sorts of risks as a loss of life expectancy compared to that of the US population as a whole. The six factors which give a loss in excess of 1000 days are: being unmarried – male; cigarette smoking – male; heart disease; being unmarried – female; being 30% overweight and being a coal-miner. Clearly the risks in life are complex social ones, little to do with technology in general or chemicals in particular. The overweight, bachelor coal-miner who smokes 30 a day and has a weak heart need not worry about any risk from synthetic chemicals. The whole question of risk is well treated in reference 17.

To summarise this chapter, I believe the main need is to emphasise once again that it is not easy to see a chemically induced health problem superimposed upon the problems normally present in the community (and often not recognised by us). We have therefore to establish the normal, then try to find the abnormal. First definitions must be made, then information has to be gathered and finally this must be subjected to rigid mathematical analysis. It is not a game for amateurs and I have not attempted to describe the methodology here. If you want to go on, consult the textbooks, such as that by Monson (5). And, once again, accept nothing without considering it carefully.

References

1. MMWR (1984). *Morbidity and Mortality Weekly Report*, **33**, No. 6, 17 February 1984, p. 75.
2. Burslem, R. W. (1984). Iconoclastic text on abortion. *British Medical Journal*, **289**, 1687.
3. MMWR (1982). Annual Summary 1981. *Morbidity and Mortality Weekly Report*, **30**, No. 54, pp. 108–9.
4. Pochin, E. E. (1974). Occupational and other fatality rates. *Community Health*, **6**, 2–13.

5. Monson, R.R. (1980). *Occupational Epidemiology*. Baton Rouge: CRC Press.
6. Connor, S. & Pearce, F. (1985). Body blow to long-term heart study. *New Scientist*, 31 January, p. 7.
7. Newell, J. (1985). New research links Agent Orange with cancers. *New Scientist*, 7 February, p. 6.
8. Enterline, P.E. (1985). Evaluating cancer clusters. *American Industrial Hygiene Association Journal*, **46**, B-10–B-13.
9. Ennis J. (1985). Statistics, St Petersburg and Sellafield. *New Scientist*, 2 May, 26–8.
10. Upton, A.C. (1982). The biological effects of low-level ionising radiation. *Scientific American*, **246**, 29–37.
11. Prescott-Clarke, P. (1982). *Public attitudes towards industrial, work-related and other risks*. London: Social and Community Planning Research.
12. Task Force (1984). *Science in the Streets*. Report of the Twentieth Century Fund Task Force on the Communication of Scientific Risk. New York: Priority Press.
13. Jenkins, R. (1985). Minor psychiatric morbidity in employed young men and women and its contribution to sickness absence. *British Journal of Industrial Medicine*, **42**, 147–54.
14. MMWR (1983). Epidemic psychogenic illness in an industrial setting – Pennsylvania. *Morbidity and Mortality Weekly Report*, **32**, 287–88 and 294.
15. Hay, A. (1982). *The Chemical Scythe: Lessons of 2,4,5-T and Dioxin*. New York: Plenum Press.
16. Cohen, B.L. & Lee, I.S. (1979). A catalog of risks. *Health Physics*, **36**, 707–22.
17. Fischoff, B., Lichtenstein, S., Slovic, P. Derby, S.L. & Keeney, R.L. (1981). *Acceptable Risk*. Cambridge University Press.

8

Cancer and chemicals: epidemic or self-inflicted wound?

During the last 100 years or so we have succeeded in suppressing or eliminating many of the mortal diseases that previously ended lives abruptly. In developed countries this trend of reduction in the death rate of juveniles and young adults has been apparent since late medieval times, but much more so since 1900. However, we have not shed our inherent mortality; death is still inevitable. The causes of death have shifted from infectious disease (particularly in childhood) to the diseases of old age, e.g. degenerative ones such as heart disease, and those forms of cancer which take a long time to become active. It appears that there is an inbuilt limit to the human life span so that with the elimination of early death from infectious disease, the main bulk of deaths are pushed down the age scale to 70 or 80 years, but not beyond (1).

The result of this change is that cancer is a much more prominent cause of death than it was formerly; i.e. it forms a higher proportion among other declining causes. Figure 8.1 illustrates the change for the US population. The risk of death not associated with cancer for persons under 65 has decreased markedly. The risk due to cancer is still not high, as you realise when you peer at the lower parts of the figure. Nor has the risk of dying from cancer increased, except for the male risk of respiratory cancer and now, perceptibly, that for females also. Nevertheless something amounting to an hysteria about cancer is now evident in the communications media, and searches for the causes of cancer are in vogue. The more detailed changes illustrated in Fig. 8.2 are valuable because they also contain incidence data (Chapter 7). The conclusion from the figure is that cancer of the lung should be our main worry. Modern changes in the environment are often blamed for the suspected increase in cancer occurrence: one suspect factor is the greatly increased use of synthetic chemicals. It is the purpose of this chapter to try and determine what influence modern chemicals and the chemical industry have had on the incidence of the various cancers.

The concept of there being just one cause for cancer is not valid. More than for any other disease, cancers seems to originate from a whole variety of circumstances. One circumstance which contributes to the formation of at least some cancers is exposure to chemicals. These may be synthetic or they may be naturally occurring. The synthetic ones are the product of the Chemical Age and are encountered in the dyestuff and rubber industries among many others, and may also be taken as drugs to treat various diseases. The naturally occurring carcinogenic chemicals occur in foodstuffs and such semi-artificial products as chimney soot and cigarette smoke. A third class of chemical in this context is the naturally occurring one which is brought to society by extractive industry, e.g. asbestos. A prime example of a synthetic carcinogen is 2-naphthylamine, a cause of bladder cancer; a natural group of carcinogens are the mycotoxins present in mouldy grain. We will consider these two chemicals in detail later but, for the moment, there is one interesting point to make.

Bladder cancer has been clearly related to exposure to certain chemicals, which resulted from a non-natural process. Control of these chemicals reduces the risk of bladder cancer. Cancer of the stomach is linked to natural carcinogens in the diet. Modern food processing and control (a product of the Chemical Age) has improved the quality of food to the point that stomach cancer is declining steadily as a cause of death (but not in Japan, see later). Thus, on the debit side of the Chemical Age, are the

Fig. 8.1 Annual age-standardised death rates, 1933–77, among Americans under 65 years of age (reproduced from reference 8 by permission of Sir Richard Doll).

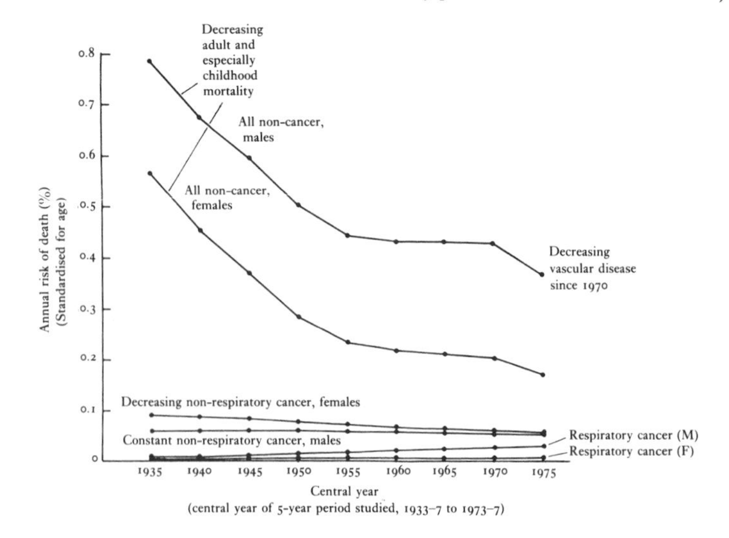

premature deaths due to exposure to 2-naphthylamine, asbestos, etc.; on the credit side is the decrease in deaths resulting from improvements in diet, hygiene, etc. It therefore cannot be said that the modern Chemical Age has resulted in an overall increase in the frequency of cancer; it has resulted in some increases and some decreases. Where the balance lies is difficult to say because the quantitative evidence is not complete and the interpretation is difficult.

Only big changes in cancer incidence can be seen in the general population, due to a variability caused by other natural factors and by difficulties in reporting. Thus for cancer death rates in males in the USA, only one type has shown an obvious increase over the last 50 years: lung cancer (Fig. 8.2). For females lung cancer has also increased, more so since about 1960. The cause of this increase in lung cancer is undoubtedly cigarette smoking. The cancer is therefore of chemical origin, but results

Fig. 8.2 Trends in the incidence of and mortality from cancer of all sites combined, and of the lung. *I*, incidence; M, mortality. Continuous lines refer to whites and broken lines to non-whites. The data are derived from seven areas of the United States. The figure is reproduced from reference 9, which should be consulted for details of how the data were assembled (reproduced by permission of Dr Devesa).

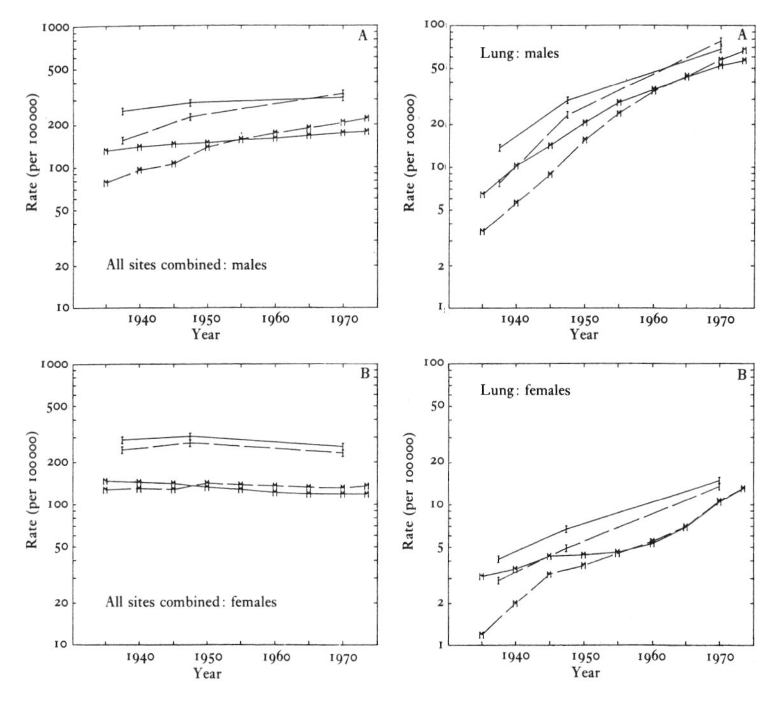

from the modification of natural chemicals by heat and is administered voluntarily.

What is therefore the meaning of public concern about chemically induced cancers, when the chief cause is a self-administered and socially accepted habit? Perhaps we are examining a social problem rather than a chemical and technical one.

It is probably easiest to discuss chemically induced cancers in terms of specific examples, rather than to consider generalities. Therefore I will discuss (*a*) bladder cancer initiated by synthetic dyestuffs chemicals, (*b*) stomach and other cancers related to natural chemicals in diet, and (*c*) lung cancer as a product of exposure to cigarette smoke and dust.

First, however, I will outline the mechanism of malignant tumour (cancer) formation as it is understood at present. This is a brief summary of a complex process which concentrates on chemical involvement in cancer formation, and ignores other possible factors such as virus involvement. The process can be considered as a two-stage one of initiation followed by promotion. Initiation involves the alteration of the genetic material (DNA) in the cell to form a mutant gene. This is a chemical reaction between the initiator (mutagen) and the DNA. The existence of the mutated DNA produces no noticeable change in the cell, and the damage done by the mutagen may in fact be repaired by the cell (see Fig. 8.3). However, if the mutated DNA is exposed to certain influences (which include chemical ones) then a process of unusual cell growth may occur (promotion) to produce a benign or malignant tumour. Tumour production thus depends primarily on the combination of initiating and promoting influences, but other factors influence the outcome. Thus a genetic lack of the DNA-repair mechanism predisposes the individual to cancer since successful mutations of DNA will persist, and conversely the presence of other chemicals (such as vitamin A) reduces the chances of promotion.

Mutagens are initiators and may also be promoters. In the latter case they are therefore also carcinogens. However, mutagens or promoters are not necessarily carcinogens, although some substances fulfil both roles and are therefore carcinogens. Where widespread initiation has occurred a promoter will also effectively be a carcinogen since there is no restriction of its opportunities to form tumours. Figure 8.3 illustrates this two-stage concept of carcinogenesis, and gives a few examples of chemicals which activate the process. The two-stage model is no doubt a gross simplification of chemical carcinogenesis, but helps to explain many observations. Thus the role of tobacco smoking as a cause of lung cancer and a factor in other cancers is explained as a general initiation in the lungs and other organs, which then greatly increases the chances of promotion by factors such as tobacco tar, dust and asbestos in the lungs, or saccharin (a suspected but not convicted carcinogen) in the bladder. Secondary factors which influence the ability of chemicals to act as mutagens or promoters relate to the activation of chemicals within the body. Environmental, food, or synthetic chemicals may not be carcinogenic when taken into the body, but become so because of chemical changes that the body carries out. Thus many potential carcinogens in foodstuffs exist as compounds with sugars (such combinations are known as glycosides). In the body the sugars are split off, releasing the active carcinogens. Conversely, carcinogenic substances taken into the body may be converted to other, harmless, chemicals.

This simple picture of carcinogenesis can be extended to take account of

Fig. 8.3 The two-stage theory of cancer induction by chemicals. The examples of mutagens and promoters given are a selected few from many suspected agents. However, very few are confirmed mutagens and promoters.

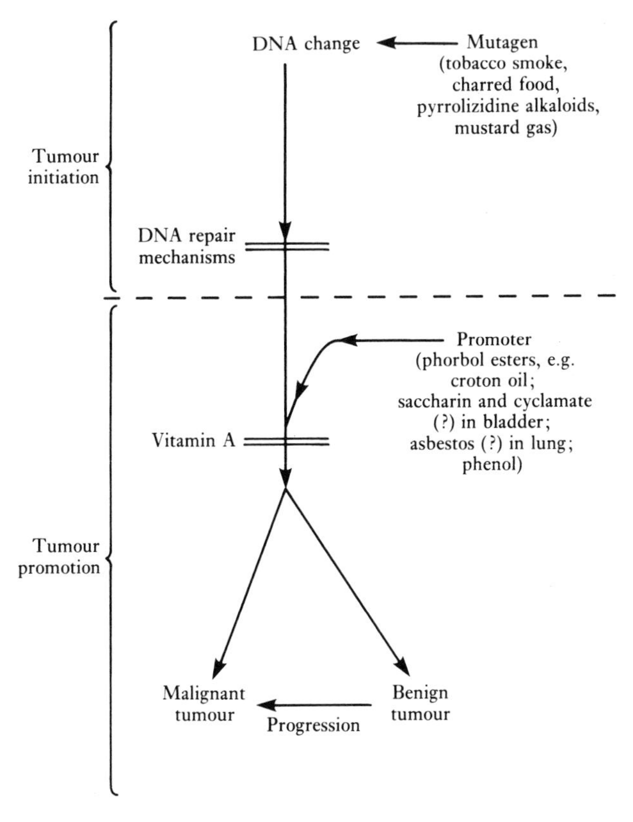

the concept of oncogenes. These are genes within the nucleus of the cell which have been modified to become potential instigators of cancerous growth by the loss of mechanisms which normally control their activity. The modification occurs either by translocation of the gene to another chromosome, or by a chemical change in the neighbourhood of the gene. Either way, the oncogene results from a change in the environment of the gene (2). Genetic material from invading viruses may also form oncogenes. Many factors, which may include chemical ones, are needed to cause an oncogene to express itself as a tumour. We do not know precisely how oncogenes fit into the two-stage theory given above; they are the physical expression of part of the initiation process, and help us to appreciate the great complexity of that process. This brief outline of cancer formation (carcinogenesis) should at best impress on the reader the fact that we can never look at just one cause of a cancer.

German skill in synthetic chemistry in the second part of the nineteenth century led to a German predominance in the production of dyestuffs and stains. This not only aided the textile industry but had stimulating effects on the German microscopists and histologists who were mapping the tissues and cells of the body with their new Zeiss microscopes, for the various dyes had selective staining effects on the different tissues and cell organelles. However, the synthetic dyestuffs were not totally beneficial for, in 1895, Rehn reported three causes of bladder cancer in workers from one factory making magenta (fuchsin) from aniline, and one case from another factory in which the patient was operating the same process. Rehn believed that aniline was the common causative factor, so that the condition became known as 'aniline cancer'. It was subsequently noted in other countries with chemical industries. In 1938 it was shown by Hueper that 2-naphthylamine caused bladder cancer in dogs, and an epidemiological study by British scientists reported in 1954 confirmed that bladder tumours in industry were associated with the manufacture of auramine and magenta from aniline, and not with the synthesis of other aniline-based dyes. Bladder cancer is thus a result of exposure over a period of time to aromatic amines containing two rings, of which benzidine, 2-naphthylamine and magenta are typical examples. 'Aniline cancer' is a misnomer. It is also of importance that the cancers may be slow to develop, and may not be apparent until some years after the victim has ceased working with the chemicals.

Although 'aniline cancer' was described in 1895, control of exposure to the causative compounds was introduced very slowly. It was in the late 1950s that I was made aware of the dangers of handling benzidine, and that this compound was withdrawn from general supply. This delay is even more regrettable because bladder cancer is a peculiarly nasty disease. Approximately 60% of patients diagnosed as having malignant bladder tumours may expect now to survive 5 years and hence probably die of another cause; conversely the mortality rate is therefore about one in two persons diagnosed as sufferers.

Naphthylamine and benzidine were used extensively as colour reagents for the identification and estimation of other chemicals in analytical laboratories and general chemical laboratories. Thus apart from the dyestuffs and rubber industries, many other chemists were exposed to them, and the bottle of benzidine I hastily discarded in 1959 was intended for analytical work. In retrospect, it is difficult to see why exposure to these chemicals was allowed for so long. Certainly the epidemiological study that confirmed the danger was reported in 1954, but 2-naphthylamine was a known carcinogen since 1938, and 'aniline cancer' has been recognised for over 50 years. The reasons for continued use seem to have been.

1) sheer ignorance (toxicological and safety information was not widely disseminated);
2) indifference on the part of the employer/manager;
3) fatalistic acceptance by the employee of occupational disease.

Whatever the reason, the current public concern and strict regulation should result in quicker action in future, at least in the laboratories and industries of an established nature which are accessible to controls. Small backyard firms and irresponsible employers will always be a problem.

The history of the aromatic amines is an example of the worst features of the chemical age. An industry-related disease was described, no action was taken, and for 50 years workers were exposed to the risk of a mortal disease. As a proportion of the total population, the numbers at risk were small, so that an increased incidence of bladder cancer would not have been observable in statistics collected from the whole population. However, the very fact that the disease was closely tied to an occupational group made its early recognition easy. More difficult problems of identification would arise for a disease that was low in incidence and not tied to a recognisable subgroup of the population. Thus when the general public is at risk from a chemically induced disease which only affects a few individuals, then this disease and its cause may easily be ignored.

When we turn to our second example of foodstuffs and stomach cancer, we immediately find problems. Many chemicals occur in food which, when fed to laboratory animals at higher concentrations, cause cancers. However, at the very low concentrations in which they occur in food there is no evidence that these chemicals cause cancer in man. Also, the cancers induced in laboratory animals are rarely in the stomach; the liver is a much more common site. Therefore the belief that chemicals in food cause stomach cancer is a presumption, reasonable but not proven.

In the USA the age-adjusted death rate for stomach cancer has fallen from about 25 per 100 000 of the population in 1930 to about 5 in 1975. This steady decline is the only obvious fall noted, except for deaths of females due to uterine cancer. Stomach cancer is still, however, an important cause of death, second only to lung cancer as a cause of cancer deaths in England and Wales in 1976, although less prominent in US statistics. Among women, breast cancer (Fig. 8.4) displaces that of the stomach in relative importance. The latter is the most common cancer in Japan. It is also a condition for which the chances of survival are low.

We know therefore that the incidence of stomach cancer has fallen steadily during the period that the production and use of synthetic chemicals has increased equally steadily and during the time that the refinement of foods has been extended. We cannot say that the former decrease is caused by the latter two increases, although there is a basis for a reasonable presumption. In fact we do not know that stomach cancer is due wholly or partly to chemical causes. It may well be that other changes in dietary habits are responsible for a decline in stomach cancer. Modern methods of food preservation (cold storage, canning, etc.) have reduced the

Fig. 8.4 A breast lump. Breast cancer is currently the commonest of the major cancers in females. The cause of this type of cancer is as yet ill-defined, but obviously complex (photograph courtesy of Peter MacCallum Hospital).

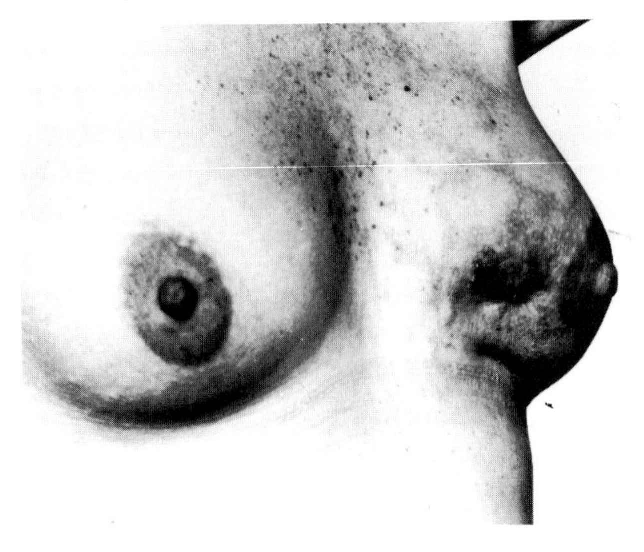

need for the traditional methods of salting, pickling and smoking. Persons of Japanese origin living in California and Hawaii have a lower incidence of stomach cancer than the inhabitants of Japan, but they have a higher incidence of cancer of the colon. This difference between populations presumably lies in their diets, as it cannot be the result of genetic differences. Another difference between modern and traditional diets is the lower proportion of roughage and fibre in the former. This decrease has been cited as a cause of colorectal cancers, as in the ethnic Japanese living in Hawaii and California. It is tempting to link the decrease in dietary roughage to the falling incidence of stomach cancer, but there is no evidence for this.

It is possible that the modern use of chemicals to prevent food spoilage has indirectly contributed to the fall in stomach cancer by displacing the traditional methods of preservation, and preventing the partial spoilage of stored foods by microorganisms.

The fact remains that chemicals do occur in foodstuffs which can be shown (at high concentrations) to cause cancers. These include polycyclic aromatic hydrocarbons (PAHs, distinct from the aromatic amines which cause bladder cancer), nitroso compounds, a variety of compounds from various plants, and fungal products resulting from the spoiling of stored foods by moulds. PAHs are present in varying concentrations in a variety of animal and vegetable products. The concentrations are increased by cooking processes which involve singeing, especially in the presence of fat, so that meat barbecued over a flame contains raised levels of PAHs. Smoked meats also contain higher amounts. Fresh vegetables and especially vegetable oils also contain the chemicals, so that it is not possible to avoid them; they are and always have been part of a 'natural' diet. They are not a result of the chemical age, although some food processing (e.g. barbecueing or smoking) may increase them. The nitroso compounds were reported to be carcinogenic in the 1950s, but the real interest in these compounds stems from an incident of food poisoning of sheep in Norway in 1962. This was caused by fish meal that had been treated with nitrite to preserve it, then fed to the sheep. A nitroso compound was found in nitrite-treated fish, and subsequently chemicals of this class were found in many meat, fish and other foods treated with nitrate or nitrite. The nitroso compounds are formed by a chemical reaction between compounds (secondary amines) which are either naturally present or formed by ageing of the food, and the nitrite or nitrate added to retard obvious spoiling. Some nitroso compounds are present in natural, untreated foods, but as yet have not been widely found. They are present in fish, sausage, bacon, ham, cheese, mushrooms and other products.

A carcinogen named cycasin has been found in the cycad nut, which is used as a foodstuff by natives of Pacific islands, notably Guam. However, the natives have recognised the toxic properties of the nut and prepare it for eating in such a way that the cycasin is removed. Therefore although cycasin is known to be carcinogenic from experiments on animals, there is no direct evidence that consumption of cycad nuts has caused cancers in man. A similar situation exists with respect to the pyrrolizidine alkaloids, which occur in a variety of plants either taken as folklore medicines (e.g. comfrey, coltsfoot) or inadvertently harvested with foodstuffs when growing as weeds in grain or fodder fields (e.g. heliotrope). Although some specimens of comfrey have been found to contain the alkaloids, it is not certain that the amounts are dangerous. Bracken fern is still eaten by Japanese people despite the fact that it has been known for some years that it contains a carcinogen. Cattle which graze on bracken-infested pastures in highland Wales and Scotland develop tumours of the bladder and, in association with a virus, the bracken diet may cause cancers in the gut of cattle. In times of drought when the cattle are forced to eat much bracken, there occurs an acute disease known as 'cattle bracken poisoning', or 'bracken staggers' if suffered by horses. Bracken seems to contain a variety of toxic chemicals with differing effects on the consumer. Many other plants in use as foods or folk medicines contain carcinogens (see also Chapter 16), and more will be found as the screening programmes continue.

Foodstuffs free of native carcinogens may be infested by moulds which produce these and other toxic compounds. The acutely toxic results of eating grain infested by fungi have been known for centuries, although the disease has not always been linked with the cause. Contamination of rye with ergot (the fungus *Claviceps purpurea*) has given rise to the disease of ergotism, which is characterised by severe mental disorder and hallucinations. The drug LSD was developed from ergot-related chemicals and used for treatment of psychiatric patients, although such use is now in disfavour. Several outbreaks of a different disease have occurred in the USSR during time of war or famine due to people eating grain that had been left in the fields over winter and had become contaminated with moulds of the genus *Fusarium*. The disease is now believed to be due to a class of chemicals called trichothecenes, of current interest in relation to 'Yellow Rain' (see Chapter 10).

It is surprising that the first connection between fungal products and cancer followed an incident of poisoning of turkeys in Britain during 1960. The birds showed both acute liver damage and liver cancers which were found to be due to a class of chemicals called aflatoxins. These were products of the mould *Aspergillus flavus* which had grown on the peanut meal supplied to the turkeys. A vast amount of research has been done since 1960 on the chemistry and carcinogenic properties of the aflatoxins. Other fungi have been found to produce carcinogenic chemicals, and the list grows steadily. One practical result is that peanuts now are examined for aflatoxin levels and many countries have an upper limit for allowable levels in peanuts before marketing. The World Health Organization and the Food and Agriculture Organization recommended an interim (as of 1966) maximum level of 30 ng/g (3 parts per hundred million) alflatoxin in human food. Thus the Food and Drug Administration (FDA) in the USA has a big programme to develop new methods of screening foodstuffs for aflatoxins and other mycotoxins. The carcinogens are not confined to peanuts, but are found in corn and many other materials that fungi can contaminate. Currently, the methods in use by the FDA can detect aflatoxin down to 100 pg/g of foodstuffs (in this case Aflatoxin B, in eggs); other toxins are detected less sensitively. This is one part in ten billion (1 in 10^{10}). Since the discovery of aflatoxins, the occurrence of fungal contamination in foodstuffs has been much more rigidly controlled, and one expects that the occurrence of liver cancer should consequently decline. It is questionable, however, whether this decline will be discernible in statistics for liver cancer mortality in future years, for many natural and artificial chemicals induce liver cancer; we do not know how important proportionately the mould carcinogens are, or were.

Various synthetic chemicals have been added to foodstuffs as preservatives, colouring agents and so forth. Regulations are becoming more stringent as to the use of these as many are suspected (with good or bad reasons) of being carcinogens. The history of the preservative AF-2 in Japan is of interest. In 1965 AF-2 was approved as a preservative for proteinaceous foods, but in 1973 it was found to be a mutagen. This provoked debate as to whether mutagenicity necessarily meant carcinogenicity, which was cut short in 1974 when AF-2 was found to cause stomach cancer in mice. It was immediately banned as an additive. The reason for its approval in 1965 can be seen as a result of insufficient testing, not through negligence but because the most appropriate and sensitive tests were not then developed. The continuing refinement of test methods means that, in future, fewer compounds will be approved, only later to be withdrawn. It should be noted that there is no evidence of any food additive, including AF-2, actually causing human cancer. This means that any cancers induced have not been in sufficient numbers to detect above the cancers arising in the population from other causes. Sugimura (3) estimates that the carcinogenic risk from AF-2 roughly equalled the risk from bracken, which is still voluntarily eaten in Japan.

To summarise this section on stomach cancer, we can find no relationship between stomach cancer and chemicals in the diet. We know there are a number of chemicals in natural diets which cause cancer (mainly in the liver). The individual will take in less and less of these as the sources of contamination are defined and eliminated. Cooking methods can be changed to reduce the formation of carcinogens in food. Known carcinogens such as AF-2 should be banned as food additives; other chemicals such as nitrites should be carefully watched and replaced in food if safer alternatives are available. At present there is no obvious substitute for nitrite. Why put nitrite in food at all? Because it kills the organism that produces botulinum toxin, one of the most poisonous substances known. The future in fact seems bright; chemical technology on balance seems to be improving the quality of food rather than spoiling it. Or rather, shall I say in relation to its disease-producing potential only? I find much of the modern processed food to be too bland and insipid for my taste.

Dust and fumes are obviously unpleasant to work in, and many work environments in the part were excessively dusty. Beyond being unpleasant such conditions were soon recognised as the cause of diseases of the respiratory system. Coal miners, quarrymen and steel workers all suffered as, to a lesser extent, did any person dwelling in a large city in which coal fires were the principal source of heat. Dust particles cause different forms of disease depending on their shape and chemical nature but, broadly, the result of this general disease (pneumoconiosis) is a stiffening or loss of elasticity of the lung and consequently a less efficient ventilation of the alveoli (the minute air sacs of the lung). Fumes, which are mainly oxides of sulphur or nitrogen, condensed with water vapour into droplets of the corresponding acids, attack the lung in a different fashion and result in bronchitis. The dust which has the most deleterious action is that from asbestos. This causes a pneumoconiosis (specifically asbestosis) which is distinct from the silicosis which plagues miners and quarrymen. This hazard from asbestos was recognised in 1930 in Britain, but not until much later was it realised that exposure to asbestos could also result in lung cancer and another cancer external to the lung. This latter cancer, mesothelioma, occurs in the body membranes lining the cavity between the internal organs and the body wall. Thus exposure to asbestos dust results in a disabling loss of lung function (which itself may be fatal in 10–20 years) followed sometimes by one of two cancer types, either of which is usually fatal very soon.

Deaths in the general population due to bronchitis used to follow regularly after the London fogs. I remember the old-type London fogs which occurred in November each year. These were so thick that one could not see the edge of the pavement nor any object more than four feet away (sorry, 1.2 metres away). The after-effect was a deposition of black particles on your overcoat, face and in your nostrils. This dirt content was the main difference from sea fogs, which are relatively clean. These fogs ceased in London around 1960, when the provisions of the Clean Air Act were progressively enforced. Bronchitis has declined as a cause of death in the UK since 1960.

We have considered diseases of the lung caused by occupational factors and by environmental factors. Now we consider a social factor which completes the evil triumvirate ruling the fate of our vital airways. Tobacco smoking is said to have been introduced to England from the New World by Sir Walter Raleigh, who was not deterred by having a bucket of water up-ended over him as he smoked. Jean Nicot, the French Ambassador to Portugal, presumably introduced the habit to France, as he had the distinction awarded to him of having the principal poison in tobacco named after him. King James I wrote a treatise against tobacco, which had little effect. The querulous Scotsman much preferred his wee drop of whisky, and presumably did not need a second drug. The relatively modern practice of widespread cigarette smoking, first by men and now by liberated (?) women has, however, caused most of the current tobacco-related disease.

There is now no doubt that cigarette smoking is one, dominant, cause of lung cancer. As mentioned earlier, this disease has increased steadily in males for 50 or more years, and now (since about 1960) in females the incidence is rising rapidly. The involvement of cigarette smoking as a contributory cause of cancers of other body regions is now being defined more closely. Here the increases in incidence are marginal; decreases in the disease due to better environmental and hygienic conditions may well be offset by any increase due to smoking, so that analysis of the statistics is complex. Figure 8.5 summarises the part played by cigarette smoking as one cause of cancer deaths in the USA at the present time (from reference 4). Cigarette smoking is the major single cause of cancer mortality, accounting for 30% of cancer deaths. This can be viewed in a broader picture when it is realised that cancer is the second most frequent cause of death in the USA, with diseases of the circulatory system the most frequent (to which also cigarette smoking is a contributory cause). The smoking of 20 cigarettes a day by an adult will shorten his life by 5 years, as near as we can calculate on evidence now available. The same is true of Britain or other advanced countries with the same social attachment to cigarettes.

We started with a discussion of dust and fumes and have digressed to cigarette smoking. The two topics are closely related, however. It is fairly clear, for example, that smoking increases the chances of contracting lung cancer after exposure to asbestos dust. It is not clear whether or not smoking also increases the risk of mesothelioma. It has been concluded by Elmes (5) that the major controllable factor now in occupational lung disease is smoking. Current working conditions in industries creating dust are now improved to the extent that the main factor contributing to lung disease is cigarette smoking.

The article by Elmes (5) presents a general picture of respiratory diseases of all types. These were second to circulatory disease as a cause of death in England and Wales in 1977, and the principal cause of sickness absence. The latter factor does not reflect any sizable contribution from lung cancer, since the course of this disease is frequently to death within 6 months. On the other hand, sickness absence is a good expression of general debility caused by a disease, which may not show in statistics of deaths. The article also reproduces figures which show that the advantages gained in controlling some diseases have been negated by the increase in lung cancer. Thus lobar pneumonia, tuberculosis and bronchitis have all declined as causes of death for men aged 55–59. These declines are due to better living conditions, control of atmospheric pollution and the development of specific chemical treatments. The increase in lung cancer deaths is very largely due to smoking.

There is public concern about an increasing incidence of cancer. That concern is justified. The cause of that increase is the prevalence of cigarette

Fig. 8.5 An assessment of the contribution of cigarette smoking to cancer deaths in the USA (modified from reference 4).

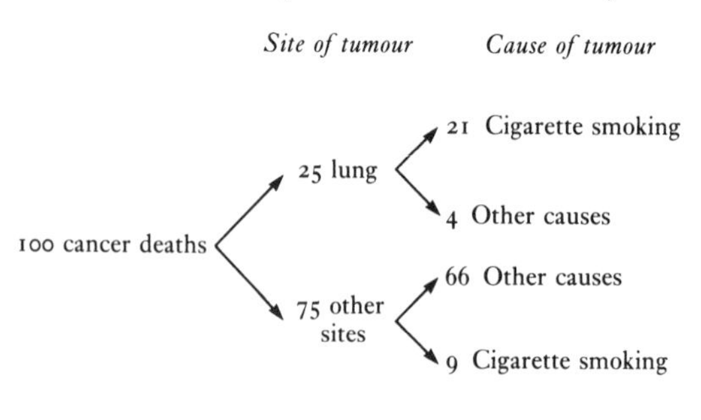

smoking, a burden that the public voluntarily assumes. Can that public then justifiably search for other causes of its disease, when the biggest cause is in every public place, every social gathering and workplace? Tobacco smoking is a vehicle for chemical exposure of the human body. The problem, however, is social, it is economic, it is political; it is not chemical. To me it is immoral to sell cigarettes, to advertise cigarettes, to collect revenue from their manufacture and sale.

The question remains that, if we sift out the complex contributions of cigarette smoking to all cancers and consider the residue, is there any indication of an increase in cancer that may be due to new chemicals recently introduced to the home, workplace or environment? The short answer is no, with three reservations. Firstly, the contribution of smoking (and to a lesser extent alcohol) as a cause of many types of cancers is still ill-defined and therefore cannot be clearly distinguished from the residual causes. Secondly, the surveys of statistics are variable and unreliable to the extent that small increases in particular cancers would be unobservable, unless the cancer was of a very rare and remarkable type (such as mesothelioma). Thirdly some cancers take a long time to develop and therefore may not be seen until long after the introduction of the responsible chemical.

Several surveys support this tentatively negative answer. The discussion by Sir Richard Doll (6) emphasises the paucity of data and the difficulties of interpretation, as does that by Peto (7), and that in their joint study (8). Better recording and interpretation of statistics will help, as will more thorough screening of new chemicals for carcinogenicity before they reach the market. I have purposely avoided a discussion of this laboratory testing; some aspects are mentioned in other chapters (see particularly Chapter 15), but the tests present great problems in interpretation. However, if we look at the various figures for incidence of cancer, the only increase obvious apart from the tobacco/alcohol-related ones is an increase in melanoma, that is cancer of the skin (Fig. 8.6). This is ascribed to a change in social habits, in which it is fashionable to expose larger areas of skin to sunlight. Ultraviolet light is a known carcinogen. Cancer of the breast in females and some other cancers appear to have risen slightly, but no explanation for this is obvious and indeed for some the rises may not be real. The difficulties of observing a chemically induced cancer epidemic are illustrated by the case of vinyl chloride. Some workers handling and using this chemical developed a rare form of liver cancer. Experimental animals developed similar tumours when exposed to vinyl chloride. The conclusion is that vinyl chloride is a carcinogen in humans; the total death toll due to this compound may be around 100 persons.

Fig. 8.6 Skin cancers at stages requiring urgent treatment. The incidence of such cancers seems to be increasing, perhaps due to the modern trend to expose oneself more to sunlight. However, education of the public has led to the increased use of protective creams, which reduces the incidence of such lesions, and also results in patients seeking treatment at an earlier stage of growth, thus improving the chances of cure (photographs courtesy of Peter MacCallum Hospital).

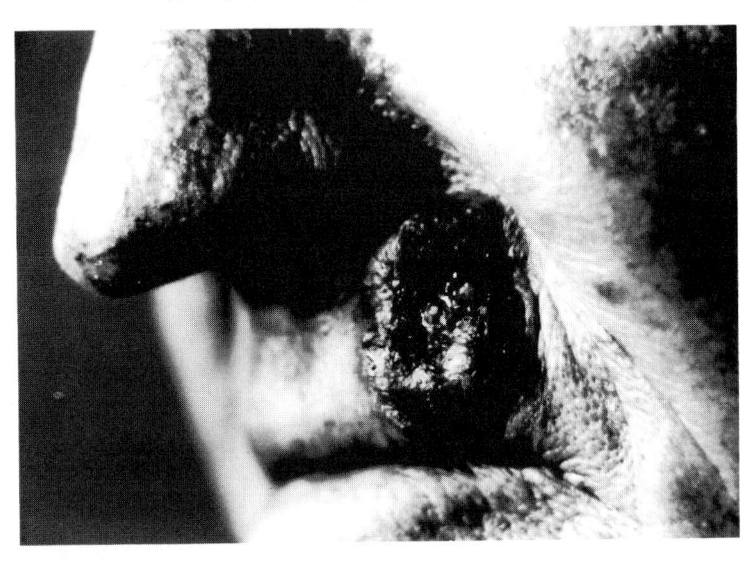

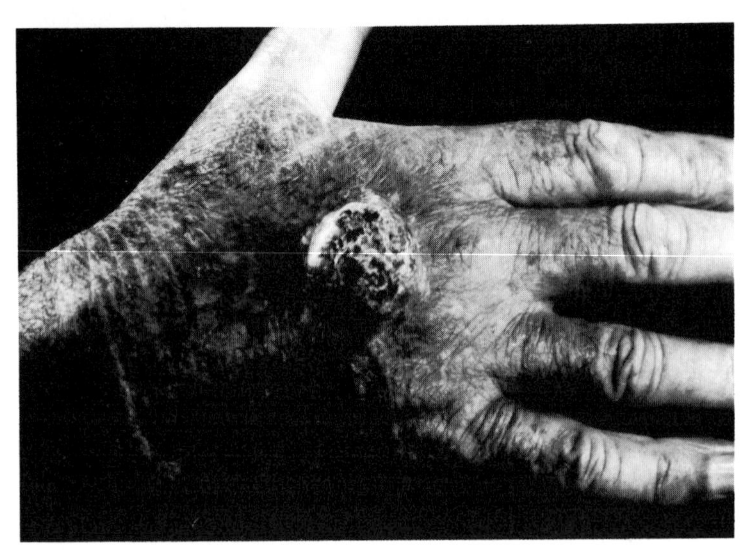

Before we end this discussion of cancer and chemicals, I must stress the distinction between the cause of the disease and its cure. We have seen (Fig. 8.2) that there has not been any large increase in the rate of occurrence of cancer in recent years, apart from lung cancer. Therefore we need not be greatly alarmed by the unfounded fear that chemicals are increasing the risk of cancer in the general population. However it is true that cancer deaths as a proportion of all deaths are increasing, simply because deaths due to other causes are declining. This is due to success in treating many diseases. Therefore it is quite right to focus more attention on research to cure cancer, for it is a major cause of death that has resisted treatment. Chemicals have not been too successful yet in treating cancer, and chemotherapy is still a fearsome process of uncertain effectiveness. We may hope that chemical treatment for cancer will become more effective, and less traumatic. Chemicals have not contributed greatly either to the causation of cancer, or to its alleviation. Our main concern about cancer and chemicals should be to improve the latter as weapons against the former.

As we look around our chemical world, we may conclude that we are safe from large epidemics of chemically induced cancers. We must, however, exercise vigilance and despite that may suffer small-scale epidemics. These may be insignificant overall but tragic to the limited populations that suffer them. We may in fact expect a reduction in cancers as our foodstuffs are better monitored for natural carcinogens, as long as we resist the current trend to exercise belief in 'organic' products, folk medicines and so forth. If we are at all honest with ourselves, we will realise that the chief chemical inducer of cancer is cigarette smoke, and perhaps in the future we will resist social and advertising pressures and shake ourselves free of this self-inflicted scourge. Or, alternatively, we will accumulate enough knowledge about all the scientific and social aspects of smoking (10) to enable us to control the habit and reduce its side effects. If this happens we will know that we are, despite all appearances, rational.

References

1. Young, J. Z. (1971). Human fertility and mortality. *An introduction to the study of Man*, Chapter 24, pp. 317–38. Oxford University Press.
2. Vines, G. (1984). Clues to cancer from a backwater of science. *New Scientist*, 1 March 1984, 10–11.
3. Sugimura, T. C. (1982). Mutagens, carcinogens, and tumour promoters in our daily food. *Cancer*, **49**, 1970–84.
4. MMWR (1982). Current trends: smoking and cancer. *Morbidity and Mortality Weekly Report*, **31**, No. 7, 77–80.

5. Elmes, P. C. (1981). Relative importance of cigarette smoking in occupational lung disease. *British Journal of Industrial Medicine*, **38**, 1–13.
6. Doll, R. (1979). National trends of disease in the post-infection era. In *Long-term Hazards from Environmental Chemicals*. London: Royal Society.
7. Peto, R. (1979). Detection of risk of cancer to man. In *Long-term Hazards from Environmental Chemicals*. London: Royal Society.
8. Doll, R. & Peto, R. (1981). *The causes of cancer*. Oxford University Press.
9. Devesa, S. S. & Silverman, D. T. (1978). Cancer incidence and mortality trends in the United States: 1935–74. *Journal of the National Cancer Institute*, **60**, 545–71.
10. Mangan, G. L. & Golding, J. F. (1984). *The Psychopharmacology of Smoking*. Cambridge University Press.

9

Drugs: the progress from natural extracts to synthetic chemicals and return

It is difficult to find a good definition of the word 'drug'. 'A substance used as a medicine' is fairly common but somehow incomplete. 'A substance recognised in an official pharmacopoeia or formulary' seems to be cheating by avoiding the issue. The popular concept of 'drug' involves the two related aspects of a medicine and of a substance which has an effect on the body although not taken purposefully as a medicine. The latter includes substances taken as stimulants, to mask feelings of fatigue or to create mental activity of a pleasant kind. These latter drugs (alcohol, caffeine, opium, etc.) are also often components of medicines. All drug substances are capable of being synthesised in the laboratory now, so there is no real distinction between natural products and synthetic drugs anymore. It is often an economic advantage for the chemist to extract the drug from a natural source and purify it, or modify it, rather than synthesise it from the beginning. Thus antibiotics are still obtained from the fermentation products of microorganisms, even though the chemist could make them from, say, petrochemicals.

A definition of the term 'drug' therefore has to encompass the two concepts of:

1) a chemical used to cure disease,
2) a chemical having a powerful effect on the body,

together with the concepts of deliberate or controlled use versus incidental presence in a product. The latter distinction is necessary as the public does not realise that many drugs are unintentionally taken along with other products. Drugs in the popular view come from a pharmacy or an illicit street-dealer; few imagine drugs are present in tea or coffee. So my definition of a drug is: a chemical used to cure disease or which has a powerful effect on the body, the administration of which may be deliberate and intentional, or unrecognised and incident upon another activity.

The source of drugs has shifted over the last century from extracts of plants and other organisms (Fig. 9.1) to chemical laboratories in which a degree of synthesis is employed along with extractive processes. It is important to remember, however, that quinine from chinchona bark is indistinguishable from quinine made in the laboratory (from simple chemicals by Woodward and Doering in 1944). The same identity holds for

Fig. 9.1 A pharmaceutical laboratory in 1840. The equipment is for the extraction of drugs from natural products, and their formulation into pills, decoctions, infusions, salves, etc. This is a reconstruction which is an exhibit of the Wellcome Museum of the History of Medicine, London (photograph: Jarrold Colour Publications, Norwich).

any drug that can be derived from both natural and synthetic sources. If we assume that the quinine from both sources is 99% pure, the difference between the two lies in the nature of the 1% impurities. These are likely to be other natural products, similar to quinine, from the chinchona bark but the impurities from the synthesis will be quite different; they will be more limited in type than those from the natural extract and will be composed of excess chemical reagents or their by-products. This question of impurities is an issue of significance we will discuss later.

The development of drugs was initiated by traditional folk medicine that was surprisingly successful in identifying the sources of useful medicines, even though it was also associated with a lot of hocus-pocus. Many modern synthetic drugs can be traced back to prototypes from natural products; digitalis drugs derived from foxglove, atropine from belladonna (deadly nightshade), tubocurarine from South American arrow poison and many others. Tubocurarine is itself isolated from the parent plant, but the scientific study of this chemical has led to the synthesis and development of novel drugs with actions similar to tubocurarine. For example, the drug alcuronium was developed by the firm of Roche after intensive studies of the mode of action of tubocurarine as a muscle relaxant. The synthetic drug offers physicians an alternative to use depending on the particular circumstance. Similar chemical variations have been made on all the other extracted drugs.

The study of naturally occurring drugs is still a source of inspiration for the design of new drugs. This study, related to the natural sources of the crude drugs, forms the science of pharmacognosy. It tends to be a minor study, pursued seriously in only a few universities. Sometimes, more serious attempts are made to learn from nature. The Roche Research Institute of Marine Pharmacology was set up in the 1970s in Australia by the big international firm of Roche in order to isolate and characterise any novel drugs that might be found in the rich marine flora and fauna of Australian coastal waters. The decision to close the Institute a few years later seems to have been a managerial one not related to the technical success of the venture.

It is perhaps worth a slight diversion to examine the natural occurrence of drugs, which will give the reader an idea of the vast number of candidates available for medicinal purposes. We will examine as an example one family of flowering plants, the Solanaceae. This family includes plants of economic importance that were of New World origin; the potato, tomato and tobacco plants. However, it also includes plants native to most parts of the world, long recognised as having medicinal or toxic properties, such as deadly nightshade (Europe) or pitchuri (Australia). Anyone familiar with tomato flowers will find it easy to recognise other members of the family. One compilation of data (1) lists 151 alkaloids as having been extracted from members of the Solanaceae. An alkaloid (alkali-like) is a basic or alkaline drug which contains nitrogen in a ring structure, occurs naturally and has some pharmacological activity. Numerically few of the 151 alkaloids listed from Solanaceae are currently used in medicine as approved drugs, but they include some very useful and important ones. Table 9.1 lists a representative selection of the more interesting drugs from members of the Solanaceae.

Of the drugs shown in the table, nicotine itself is not used much nowadays in pharmacy, although historically it was of great interest. Most modern medicinal use is made of the atropine-scopolamine family, which has been expanded with the addition of many synthetic variants.

The alkaloids of Solanaceae are a small fraction of those known; the

Table 9.1. *Some of the active drugs from plants of the family Solanaceae*

Drug	Genus[a]	Common name	Comments
Nicotine	*Nicotiana* and others	Tobacco plant	A social drug; not used medicinally, but has veterinary and horticultural applications
Solanine	*Solanum*	Woody nightshade Black nightshade Potato	Not used in medicine; occasionally people become poisoned by solanine as a result of eating green potatoes
Atropine	*Atropa* *Datura* *Duboisia*	Deadly nightshade Thornapple	Archetype of a group of drugs which reduce the effects of stimulation of the parasympathetic nervous system; an antidote to nerve gas poisoning
Scopolamine (hyoscine)	*Atropa* *Datura* *Duboisia* *Scopolia*		Closely related to atropine; similar medical uses to that drug; given to prevent motion sickness, among many other applications
Capsaicin	*Capsicum*	Peppers and capsicums	The active flavouring principle of peppers, chillies, paprika etc.; minor medicinal uses

[a] A botanical genus contains many species; the potato (*Solanum tuberosum*) is quite distinct from woody nightshade (*Solanum dulcamara*).

family Apocyanaceae has yielded 765 alkaloids (1). Add to this the alkaloids found in fungi, algae, animals and microorganisms, then add the non-alkaloid drugs also found plentifully in nature, for example the glycosides from *Digitalis* (foxglove). There is therefore an almost infinite supply of natural candidates for the pharmacy shelf and the medicine cupboard.

What I have done above is to set the background to this chapter. We come now to the purpose of all this, which is to discuss the reasons why, if synthetic chemicals are identical to natural drugs, the public is uneasy about the synthetic ones and, in fact, there is a general reversion back to herbal or natural medicines. Why do many people prefer the herbal medicine to the chemist's prescription? Is there a reaction against the scientific basis of medicine, a movement towards the irrational and, if so, why has this occurred and is it justified?

It is in fact very difficult to know if there is an increasing movement away from prescription medicines, as there is no quantitative information. What seems clear, however, is that there is no decrease in public acceptance of non-official remedies, which means that the public has not become clearly convinced of the superiority of the scientific remedy, and still seeks an alternative. The reasons for the failure of the prescription drug to conquer the opposition seem to be to some degree linked to the pressures on medical practitioners. If a patient presents himself with a minor complaint the doctor can say it will resolve itself in time. The patient is not happy with this; he or she expects a prescription, therefore it is easier for the doctor to give one. Alternatively, if a patient presents with a disorder the doctor knows is due to anxiety, it is easier to prescribe a tranquillising drug than to have to try and to re-order the patient's complex social and family life to relieve the actual causes of the anxiety. Thus, in these marginal cases, a drug is prescribed unwillingly by the doctor because either the patient expects it or it is the easiest course of action in an essentially remedy-less situation. The conscientious doctor should insist that the patient return at the end of the course of medication so that the result can be evaluated. Many doctors do not insist on this; further, many patients do not see why they should pay additional consultation fees if they are cured. Therefore, if the patient does not return, the doctor does not know if:

1) the patient was cured;
2) the medicine had no effect;
3) the patient actually took the medicine;
4) the medicine caused unpleasant side-reactions.

Absence of information on the first two means that the doctor has no feedback to guide her in future prescriptions. Information on the third is of course vital in assessing the usefulness of prescriptions. I recently read of a case in which several kilograms of drugs were recovered from a patient. These were obviously not the proceeds of one prescription, so one wonders on what the practitioner or practitioners based their assessment of the patient's progress. In a related incident, I was in hospital some years ago for the removal from my legs of varicose veins. Each night the nurse gave every patient a laxative and a sleeping pill. As I did not need either of them I kept them. When I left hospital I gave the sister the handful of pills I had accumulated. She no doubt thought my action strange; not half as strange as I found a system which firstly gave unnecessary medicines and then made no check to see if they were actually taken. Failures in 'compliance' of this kind are the result of poor communication between the doctor, the dispensing agencies (pharmacy, nurse etc.) and the patient.

Some kind of unpleasant side reaction (fourth case, listed above) is common enough with many medicines. The possibility of an adverse drug reaction is one of the factors which has to be assessed when a prescription is written; does the seriousness of the disease warrant the risk of a reaction? Such reactions can vary from mildly disturbing (nausea, headache, slight skin rash) through more pronounced and dramatic symptoms to effects which are life-threatening (a full-scale allergic reaction). If a patient suffers such a reaction, it will reduce his confidence in the prescription and the prescriber.

Antagonism of the patient to the doctor and to the prescribed medicine can therefore arise from ineffectiveness of the medicine, dissatisfaction with the prescription and hence failure to take it, or the occurrence of an unpleasant reaction or side-effect when the medicine was taken.

Such events of course only occur occasionally. In many cases the patient comes to the doctor with a disease that the latter can diagnose rapidly. The patient then receives a prescription which quickly and undramatically cures the illness with no further problems. It is perhaps the very success of this process which makes the patient fail to appreciate the benefits of modern medicine. The patient expects to be cured; failure is now the exception which creates comment. The patient with a septic cut finger does not expect to die of gas gangrene because she knows antibiotics work; she does not expect to die of tetanus because chemical antiseptics will sterilise the wound and because she had her tetanus shots at work. The father of the baby with a respiratory infection does not now dread pneumonia because he expects the antibiotics to cure the infection before the lungs are invaded. The very great successes achieved by the intelligent medical use of modern synthetic chemicals are regarded as commonplace. It is only failure that is remarked on and spread abroad.

In the wider sphere of human activities covering technologies other than medicine, there is evidence of a swing away from the rational towards an obscurantist outlook. This antipathy to the scientific outlook probably arises from a feeling that control of everyday life has been entrusted to a few scientists and technicians whose activities are incomprehensible to the general public. Or, alternatively, perhaps the public just does not discriminate between science and the occult, each being equally obscure, and therefore chooses whichever approach is more attractive for quite personal and mundane reasons. We will discuss the subject further, later on in this book (Chapter 17).

Another reason for a patient to turn away from conventional medicine to the less rational alternatives is the personal relationship between doctor and patient. In former times the physician was trained to listen to the patient and to express some understanding of the latter's condition, even if this was not a medical problem. Thus it was taught that if one listened carefully to the patient, the constituent elements of a diagnosis would be revealed, which would then only require an intelligent interpretation by the physician. Similarly, emphasis was placed on cultivation of a good 'bedside manner'. Now the diagnostic tool is the pathological test; the doctor wants the sample of urine with the minimum of personal history and (largely) irrelevant misfortunes of the patient. It is unfortunate that some of the recent graduates of medical schools are arrogant; that is the only word that really fits. A patient accorded this treatment will turn more readily to the practitioner of alternative medicine, if the latter individual is willing to be attentive and pleasant. In fairness to the medical profession, I also point out the quite extraordinary attitudes of some patients towards doctors, which has resulted in threats of violence if prescriptions are not given or repeated. As a consequence, the real advances in chemistry and pharmacology may be lost essentially by a failure in personal relationships.

Therefore, people for one reason or another continue to seek for alternatives; there is a wide range of choice. However, what we are basically offered is something which is 'natural' in origin, coupled with a philosophy of use which claims some justification in a mystical or obscure body of knowledge not readily accessible to the layperson. These philosophies vary from those with some degree of intellectual justification to those claiming origin in a purely mystical or 'given' way, with no rational basis at all. The latter usually claim a derivation from some Eastern philosophy; whether genuine or not is often hard to determine. We will first, however, consider the 'natural' concepts of alternative medicine before becoming involved in the philosophies of use.

'Natural' seems to mean a preparation that has had the minimum of processing, so that it is readily identifiable as dried leaves, root or fruit.

Some degree of formulation may have occurred, as for ointments, toiletry products etc. In still other cases, the term 'natural' may appear on the label simply because it is the fashionable selling point. What is definitely not implied is that the active ingredient is present in purified form and known concentration. Herein lies the whole problem. The practice of pharmacy arose because it became recognised that unless the purity and concentration of the active ingredient were known, the doctor could not predict the patient's response. Too much, and the patient was killed by the remedy. Too little, and the doctor was shown to be incompetent.

William Withering in *An Account of the Foxglove and some of its Medical Uses* (1785) expounds upon the problem of getting an accurate dose from a plant product. His solution (quoted in reference 2) was to gather the leaves of the foxglove only when it was flowering, to dry them carefully and then to prepare the infusion or powder. The pharmacist (Fig. 9.2) took over this function of preparation, and very gradually the methods of assaying the active principles were improved. At first these methods had to be based on the reactions of the patients or of test animals. Crude chemical tests were then developed and refined, so that for example, alkaloids were evaluated as

Fig. 9.2 Mr Gibson's Pharmacy in 1905. Many of the items for sale are patent medicines which are attractively packaged to attract sales. However, the contents come from sources not too different from those of Fig. 9.1 (the reconstruction is from the same source as Fig. 9.1).

alkalis by extraction from the crude material, then titration with standard acid solution. These 'wet' chemical methods are being increasingly displaced by instrumental techniques such as ultraviolet or infrared light spectrometry. The identification and estimation of approved drugs are included in the various pharmacopoeias, which are invested with regulatory authority. The pharmacopoeia is therefore the pharmacist's reference by which he can evaluate whether or not a drug preparation conforms to the required standard.

We have therefore progressed away from natural medicine in a simple endeavour to make treatment safer. The modern medicine has exactly the same active component as the natural medicine, but in a precisely measured amount and free from other impurities which may well have an unpredictable and confusing effect on the patient. The current movement towards natural medicines is moving the wheel back to where we were 200 years ago, before William Withering. In order to understand this, we will have to look more carefully at the successes and failures of the modern pharmaceutical drug industry, and those of the alternative medicine culture.

In Chapter 8 there is a brief account of the changes in life expectancy over the last 100 years. The increase since 1900 is due to a complex number of factors which are certainly not all chemical. Better housing and social conditions, better hygiene (in part due to chemical disinfectants), and improved nutrition (in more recent times partly due to pesticides) have all been involved in this change. It is noteworthy that this change occurred before the introduction of antibiotics, in the 1930s for sulpha drugs and 1940s for penicillin, but there is no doubt that the latter chemicals have reinforced the trend to greater life expectancy. The development of these compounds is well described by Albert (3). The pharmaceutical industry has thus helped to extend life expectancy of three score years and ten, but cannot claim sole credit for this. Many other factors have helped. Apart from the increase in life expectancy, the drug developments of recent years have helped to improve the quality of life for those suffering from debilitating but not necessarily fatal diseases. The control of blood pressure is one example which is worth examining.

The natural control of blood pressure in the body is very complex. Some of the factors involved are:

1) the mechanical condition of the pump (the heart);
2) the control systems for the heart;
3) the physical condition of the tubing into which the heart pumps the blood;

4) the mechanisms which control the flow resistance of the tubing, i.e. the degree of contraction or expansion, and the shunt systems which control where the blood goes (e.g. to the skin when the body is hot, to the intestines after a meal);
5) factors which control the volume of blood in the system, which include the distribution of water between blood and tissue, the retention of water by the kidneys, and the distribution of salt between blood, tissues and kidneys.

Factors 1 and 3 are controlled to some extent by the age of the individual and are also susceptible to diet; deposition of cholesterol inside the blood vessels is a chronic problem. Diet also has some bearing on factor 5, for a high salt diet will compound hypertension problems. It is the control mechanisms (factors 2, 4 and 5) which are capable of being regulated by drugs. In the last 20 years or so a number of drugs have been developed which act in various ways on these systems; five or six classes of such drugs can be defined, based on actions on the brain, on various parts of the adrenergic or cholinergic nervous system, on water and salt distribution, and as inhibitors of the enzymes which make the natural control factors. None of these drugs is specific in effect, and all have side-effects. To control a severe case of hypertension it is usually necessary for the patient to enter hospital for a few days while the physician works out a regime of drugs that just balances the blood pressure at a desirable level, with the minimum of adverse effects. The patient can then leave hospital, but may still experience some unpleasant effects of the continuing drug treatment. Change in this regime may be necessary later, as the patient's control systems become further impaired or, in some cases, become habituated to the drugs and therefore less responsive. It is easy to see, therefore, that from the patient's point of view such drug therapy is not ideal. Other things must be considered. A study in Australia (4) has shown that there is a proportion of people in the population that have mild hypertension, without obvious symptoms. Further, it was shown that treatment of this condition reduced mortality and morbidity in some of these people (males over 50 years old). Fine, we will treat them all. However, in terms of quality of life, this may not be the best action to take. Some antihypertensive drugs cause impotence; is a man's quality of life improved by a slight reduction in morbidity which is linked to impotence?

The benefits of antihypertensive drugs are real, when administered carefully and when account is taken of all the factors in the patient's life which add to its quality. Success is the control of a disease that can kill or incapacitate quickly. Failure is the infliction of burdensome side-effects on

Fig. 9.3 The high cost of modern orthodox medicine is illustrated by these two hospital scenes. (*a*) A patient is being prepared for computerised tomography (CT scanning), a new and effective diagnostic tool; (*b*) the patient is receiving vital cell components of blood. In both instances high capital costs combine with skilled technical time to make treatment very expensive (photographs courtesy of Peter MacCallum Hospital).

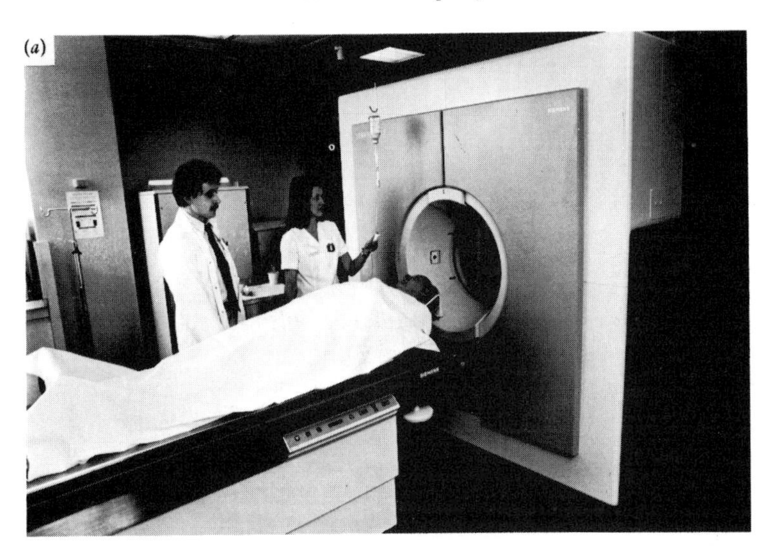

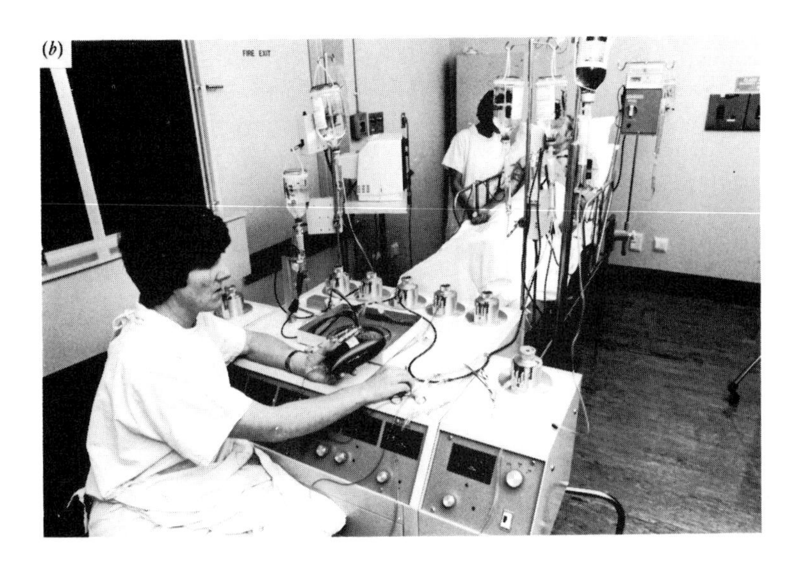

a patient who may, or may not, be better off with some form of care other than drug treatment. The public's views of the drug industry and the medical profession must be coloured by such considerations.

A further problem is that of the drug which is truly useful in a limited way, but the use of which has been wildly over-extended to a point bordering on abuse. The prime example of this is Valium (diazepam), which is also the all-time money spinner for a drug company. This chemical has clear benefits in some disease states, for instance in the treatment of epileptic attacks. However, Valium is most widely prescribed as a tranquilliser, for the relief of anxiety. It is symptomatic only, that is, the effects last only as long as the drug is taken; there is no long-term cure. It is therefore suited for management of a crisis, as a stopgap until the causes of the anxiety can be alleviated by other means. Unfortunately, because Valium is effective and is free from major side-effects or adverse reactions, there is a tendency to keep on taking it, for to do so is the easiest solution. The problem then is that the patient becomes habituated to Valium, and an increase of dose may not be effective, in fact sometimes increases anxiety. If the patient has been taking the drug for years, great crises of anxiety may be precipitated by its withdrawal. The patient has to be carefully managed over a long period whilst the drug is being withdrawn. The short-term effectiveness of the drug is thus often negated by useless long-term prescription. It is not an antidepressant, although it is sometimes prescribed as such with potentially dangerous results. Valium may be used as a substitute for alcohol by persons under stress, being more popular with females than males (who tend to prefer alcohol). If the drug is denied to a patient, there is therefore a risk that he/she will turn to alcohol as a substitute, or to other drastic means of alleviating the stress. What is often necessary is a radical change in the patient's life style, something very difficult to arrange. The argument against Valium is therefore concerned with the sheer volume of the drug that is used.

The topic of the apparently high profits of the drug industry and its attempts to promote drug sales by all possible means is the subject of much discussion. The book by Klass (5) summarises the concerns that are still current. Here, I will only stress the need to consider all factors objectively. Drugs of proven value must be available to the community at reasonable prices, and all information about them and their effects must be made public as soon as it is known. In return for developing and supplying the drugs, the industry must receive payments that realistically reflect the costs of development and production. As standards of safety increase, so do costs of development, and therefore the community may expect to pay more for a new drug. What is a reasonable reward to industry will always be a matter of debate, but it is imperative that the facts be known.

So much for the concerns about the synthetic drug industry. What can alternative medicine offer us? Many forms of this are non-chemical, e.g. various types of massage, acupuncture, faith healing, pattern therapy, pyramid healing and yet odder remedies (see reference 6 for a list of the alternative medicines). The chemical forms are divisible into two classes: herbal and non-herbal. The herbal medicines include the mainstream Western tradition (7) and medicines from many other cultures: Unani and Ayurvedic from India (8), traditional Chinese (9), current US folk medicine (10), Australian wild medicine (11). There are also more recent derivatives of herbal medicine, such as aromatherapy (6). These are the traditions that led to modern pharmaceutical science as outlined above.

The non-herbal alternative medicines include homeopathy, biochemics and megavitamin therapy. Homeopathy cannot be explained in current scientific terms (9) and attempts to give it a logical explanation (6) are entirely unsupported by fact. It is, however, not a dangerous practice in itself as the medicine is given in such low dilutions as to be without any effect, bad or good. Biochemics is claimed to cure by restoring the balance of inorganic salts in the body by appropriate dosing with (apparently arbitrary) combinations of salts. Like homeopathy it is ineffective but relatively harmless. This cannot be said for megavitamin therapy. The principle of this treatment seems to be that if a little of something is good for you, then a lot is much, much better. This somewhat naive concept is contrary to the definition of a vitamin, which is a dietary requirement in small amounts not forming a substantial part of the bulk of the diet. It is also contrary to the concept of appropriate dosing (see Chapter 4), which (to their credit) is emphasised by the practitioners of other branches of alternative medicine.

High doses of particular vitamins do have therapeutic value in a few defined diseases (12). Very high doses of vitamins to treat ill-defined diseases either have no effect (if the patient is lucky) or are positively harmful. Toxic effects of megadoses of vitamins A, D and B_6 are known to occur (13). It would not be surprising if other vitamins are in future shown to have harmful effects at high and continuous doses. The authors just cited also make the interesting observation that impurities in vitamins become important when the dose is elevated. The US Food and Drug Administration demands that a synthetic vitamin be at least 98% pure, and preclinical testing must show that a 2% impurity is not toxic at normal, recommended doses of the vitamin. What happens if the individual then takes 1000 times the normal dose? He is in a potential danger zone as regards dosage. The worst aspect of this is that the chemical nature of the 2% contaminant is unknown. At least the potential toxicity of the pure vitamin can be assessed as the chemical nature is defined. Contaminants can vary from one manufacturing batch to another. Toxicity resulting from megavitamin dosage may be very difficult to link to a specific contaminant, given that it would take at least 6 months for the toxic effects to be described and analysed and, in that time, the responsible batch of vitamin may have been consumed or dissipated into the market. In Australia (at least in the State of Victoria) there is no control over the marketing of a vitamin preparation, provided that no specific claims are made for the preparation. The same probably applies to many other countries, less active than the USA in regulatory matters.

I find it very hard to understand a public that will happily eat large quantities of partly unknown chemicals, yet become obsessively agitated by trace contaminants of a toxic chemical (dioxin) in a herbicide not taken by mouth and not shown to have any deleterious effect on people when used as intended (see Chapter 13). The philosophy of megavitamin therapy seems to ignore the principles of effective dosage and of balance between beneficial and harmful results. Vitamins are essential at low doses, they may be beneficial at higher dose levels for a small number of defined disease states, but some can also be toxic at chronic high doses. Unfortunately the public is getting the impression that the more vitamins they take, the better their health will be. Proponents of the therapy may indeed caution their patients about indiscriminate dosing, but the general public can buy and consume as many vitamins as they wish without restraint.

I will now try to analyse more fully the public attitudes to alternative medicine (of the chemical type). Herbal medicine has been fairly fully discussed previously. There is a scientific justification for this practice, with the reservations mentioned above about potency and purity. When one looks through older (7) or recent (11) books on herbal remedies, the impression is that the useful information is diluted by blanket claims that one particular herb will cure everything, so that the actual, useful qualities are obscured. Nevertheless, there is ample, objective evidence for the effectiveness of the active principles of the herbs, e.g. digitalis from foxglove, atropine from deadly nightshade, etc. This question of objective proof is the key to the public attitude to alternative medicine. In short, the public does not require proof, therefore it will accept almost any proposition that is well presented. The proponents of alternative medicine do not always feel that proof of the beneficial effects of the therapy is necessary. Stanway (6) argues that, for various reasons, controlled trials are difficult. No doubt they are, but they should be attempted, and must be done if the treatment is to be fully accepted. A controlled trial is one in which the subjects and actual experimenter are ignorant of which treatment is the one under test, and which is the control (i.e. the point of comparison). The observations are collected and analysed before the nature of the treatment is revealed. This removes any bias from the results, at least it should do if all goes well. Various problems can arise, due to faulty procedures, or judging by the investigator (14). However, if the trial is properly recorded the conclusions can be reassessed later if it is felt that the original conclusions are in error. This kind of trial establishes a practical basis for claims about a drug.

The proponent of alternative medicine relies on two phenomena to 'prove' his claims. One is that many minor maladies will resolve themselves. Therefore success for a treatment is assured in a certain proportion of cases. Secondly, a certain amount of sympathetic attention from the practitioner will make the patient feel better, whether or not there is any physical improvement in his condition. I do not decry this psychological reassurance as being useless, or fraudulent. It is a part of medicine, in fact it was the original basis of all healing. We should, however, be clear about what is being done and achieved. If the patient is reassured to the point that he will more positively set about improving his own condition, well and good. But if it is tied to a pseudoscientific mumbo-jumbo that involves chemical treatment that is harmful, then we should object to it. Thus I prefer homeopathy to megavitamin therapy. The alternative medicine practitioner will always be able to point to a certain proportion of successful treatments. What must be proven is that these improvements in health would not have occurred if the treatment had not been given, and that the operative element in real cases of success was not just psychological (important though that may be). These practitioners seems to have all sorts of reasons why such proof cannot be given in their case.

The two big worries about alternative chemical medicines are firstly that they may do actual harm to the patient by administering toxic chemicals at dangerous doses and secondly that the patient may be diverted from successful orthodox treatments.

The reasons for acceptance of alternative medicines are illogical, as demonstrated above. The public just is not interested in logical argument or proof. There is much more interest in something invested with an air of mystery; the occult is preferred to the known. The plain, dry facts of orthodox medicine, argued with statistics and adorned with incomprehensible terminology, have no appeal. The doctor and pharmacist have to perform with a white coat as sole uniform. Perhaps they would do better jazzed up a bit with the trappings of shamanism and witchcraft, with a bit of necromancy thrown in. Come back, Dr Faustus! We must conclude that people are not logical creatures.

I have not discussed in detail the efforts of the drug industry itself to promote its synthetic products (5). Undoubtedly, a reaction against what is seen as an unscrupulous and powerful pressure group accounts for some of the popularity of alternative medicines. This may be yet another illogical action. Is there any reason to believe that the manufacturers and distributors of so-called natural remedies are any less animated by the profit motive or are more scrupulous than the manufacturers of synthetic, registered drugs? For all the public knows, the two types of manufacture may be owned by the same corporation. At least the drug corporation is subject to more formal regulation which, however, it may do its utmost to avoid, if Stanley Adams' experiences (15) are typical.

Further illogic is exhibited in the attitude to social drugs. The natural chemicals have always been with us. In western European culture the drug has traditionally been alcohol, to which were added nicotine and caffeine during the expansion of trade in the sixteenth century. Afterwards came opium, hashish and latterly cocaine. All are damaging, both in terms of social disruption and in terms of toxic effects to the consumer. Alcohol and tobacco increase the risks of cancer (Chapter 8), and recently it has been found that cocaine, the usage of which has now become epidemic, causes damage to the liver of those taking the drug (16). Yet it is those persons who are most critical of conventional, medicinal drugs who seem most likely to dabble in these social drugs as part of the alternative culture. I should except alcohol, caffeine and nicotine from the preceding statement, as these three are established drugs. Am I then saying that these three are acceptable, and other drugs are not? No. We have enough trouble dealing with the social problems of alcohol alone. If we can discourage the acceptance of further drugs, we should do so. Hypocrisy is important in attitudes to social drugs; we may have to accept provisionally those that are socially entrenched (and risk being hypocritical), but recognise that all are potentially evil without adequate control. Therefore I do not see the logic of embracing novel drugs as a reaction to older social drugs and to the synthetic, medicinal drugs.

Synthetic drugs have developed from herbal medicine and other chemical origins to become a substantial element of the chemical world. Natural and synthetic drugs do not differ, but society still has trouble accepting this. The synthetic drugs have become the tools of orthodox medicine, introducing problems of their own. Many of the perceived failings of modern medicines are due to communication problems between industry, regulatory authorities, practitioners and patients, rather than inherent chemical problems. The failings in the orthodox system have led to renewed interest in traditional (Fig. 9.4) or modern alternative systems of medicine, some of which are not harmless. The arguments used to support alternative systems are based on hypotheses not susceptible to proof, but society is willing to accept this because its nature contains an irrational element. The use of cold logic is unlikely to rehabilitate the orthodox system, but a more human approach to the patient within it, and the correction of the more obvious abuses, will certainly help. These are, then, the conclusions of this chapter.

In the section of the chemical world devoted to drugs, people stand uncertainly between a rational system they do not understand, and various irrational systems which are superficially attractive. Collect the facts before you decide which to choose.

Fig. 9.4 The bent arrow, or how to get back to where you started from 200 years ago. A widespread acceptance of herbal and alternative medicine will reject two centuries of progress in pharmacy and pharmacology.

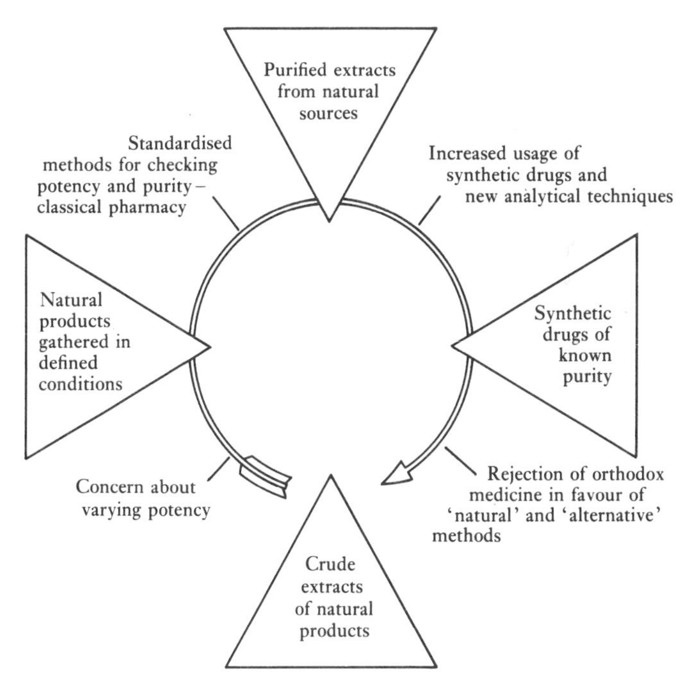

References

1. Raffauf, R. F. (1970). *A Handbook of Alkaloids and Alkaloid-containing Plants*. New York: Wiley-Interscience.
2. Holmstedt, B. & Liljestrand, G. (1963). *Readings in Pharmacology*. Oxford: Pergamon.
3. Albert, A. (1965). *Selective Toxicity*. London: Methuen.
4. Management Committee (R. Reader, Chairman) (1980). The Australian therapeutic trial in mild hypertension. *The Lancet*, June 14, 1980, 1261–7.
5. Klass, A. (1975). *There's Gold in Them Thar Pills*. Harmondsworth: Penguin.
6. Stanway, A. (1979). *Alternative Medicine*. Adelaide: Rigby; London: Rainbird.
7. Culpeper, N. (1826). *Culpeper's Complete Herbal and English Physician*. Manchester: Gleave. This 1826 edition by an anonymous editor has been reproduced in facsimile by Harvey Sales (London?) in 1981.
8. Aslam, M. (1983). Asian Medicine in Britain. In *Pharmacognosy*, Trease & Evans (1983), *op. cit.*, pp. 666–80.
9. Trease, G. E. & Evans, W.C. (1983). *Pharmacognosy*. London: Ballière Tindall.
10. Aikman, L. (1977). *Nature's Healing Arts: From Folk Medicine to Modern Drugs*. Washington: National Geographic.
11. Cribb, A. B. & Cribb, J. W. (1981). *Wild Medicine in Australia*. Sydney: Collins.
12. Ovesen, L. (1984). Vitamin therapy in the absence of obvious deficiency. What is the evidence? *Drugs*, **27**, 148–70.
13. Rudman, D. & Williams, P. J. (1983). Megadose vitamins, use and misuse. *The New England Journal of Medicine*, **309**, 488–90.
14. Barber, T. X. (1976). *Pitfalls in Human Research: Ten Pivotal Points*. New York: Pergamon.
15. Adams, S. (1984). *Roche versus Adams*. London: Jonathan Cape.
16. Kloss, M. W., Rosen, G. M. & Rauckman, E. J. (1984). Cocaine-mediated hepatotoxicity, a critical review. *Biochemical Pharmacology*, **33**, 169–73.

10

Chemical warfare and disarmament

Many of the chapters in this book concern the use by man of chemicals to attack other species of living organisms, e.g. antibiotics against bacteria, herbicides against plants and insecticides against insects. This interspecific chemical war is a continuing battle consuming large quantities of synthetic chemicals and, in economic terms, is a major support for the chemical industry. We have also used chemicals in wars against our own species, the intraspecific chemical war. This has fortunately not been continuing or intense, except for one short period, but is certainly of significance to us in our view of the chemical world. In the following account I have tried to keep to an objective summary, and avoid discussions of the moral implications. I am not a proponent of chemical war or of war in general, but it is a subject that must be discussed to complete a picture of society in the chemical age.

The use of toxic chemicals in warfare has a long history but, until 1914, they were not used on a large scale. The use of curare to poison arrow tips, or the poisoning of wells are activities which could be described as chemical warfare, but the First World War produced a completely different type of activity in which large quantities of synthetic chemicals were used. Strangely enough, this was also the last occasion in which chemical warfare was practised on such a scale. The reason for the use of chemicals at that time was probably because the chemical industries of western Europe had just completed their first big expansion, so that toxic industrial chemicals were available in quantity.

The German army is believed to have used tear gas in late 1914 or early 1915, the first reported use being an attack on the Russians at Bolimow with artillery shells containing xylylbromide on 31 January 1915. However, the chemical war really started on 22 April 1915 when the Germans released 180 tons of chlorine from 6000 cylinders on a front of 6 km near Ypres. This cloud drifted onto the opposing British and French forces, who were totally unprepared. There were 15 000 casualties, of whom

5000 died. Further massive quantities of chlorine followed; 720 tons were used in the Champagne for an attack in the autumn of 1915. The British and French hastily improvised crude respirators to protect against chlorine. As this gas became less effective the Germans used phosgene (December 1915) and various irritant or sneezing chemicals alone or in combination. The situation changed markedly on 12 July 1917 at Ypres when the Germans first used sulphur mustard. This chemical has a toxicity far in excess of anything used previously, yet the effects are insidious and do not appear until several hours after exposure. Liquid mustard on the skin causes burns or blisters, as does the vapour after prolonged contact, and blindness if it gets in the eyes. When inhaled as vapour or fine aerosolised liquid, mustard causes destruction of the respiratory tract. Mustard at high dosage is lethal, but the main result of exposure is to render the victim temporarily or permanently unfit to fight. It is a casualty-producing substance of great potency and persists in the environment for long periods, contaminating and penetrating most objects it contacts. The Allies did not make mustard until late in the war; the usage was almost entirely German.

After the initial use of chlorine by the Germans at Ypres, all combatants felt free to use chemicals. Towards the end of the war a pattern emerged of mustard gas employed in shells by the Germans, phosgene in clouds or projected drums liberated by the British and hydrocyanic acid (prussic acid) in shells used by the French. The US started a large-scale manufacture of Lewisite, but this was too late for use in the war. It is estimated (1) that 125 000 tons of all types of battle gases were used in the war, causing 1.3 million casualties. This estimate is very approximate; the British claim (2) to have suffered 181 000 casualties including 6109 dead, and to have delivered 5700 tons of gas to cause approximately 170 000 casualties to the Germans.

The result of all this activity was to produce a general burden of casualties and deaths on both sides; in this war of attrition neither side could claim that chemicals helped to resolve the conflict, but they certainly helped the process of attrition in an impartial manner. Chlorine and mustard produced debilitating injuries to the lungs, whereas phosgene caused a high proportion of deaths. This latter effect was due to secondary infection of the damaged lungs by bacteria; no antibiotics were then available to treat this. The Russians are said to have suffered particularly from chemical weapons, despite their withdrawal from the war in 1917. Perhaps this relates to an inability to produce the defensive equipment that the Allies made.

At the close of the war, three and a half years of intensive effort had produced a vast technical knowledge on the offensive use of, and protection against, war chemicals. Sulphur mustard was undoubtedly the most potent chemical agent available. The technology of protection had advanced very considerably and today is different only in detail. Many people at the end of 1918 were convinced that future wars would be fought with chemical weapons. This led to a consideration of disarmament proposals. The Hague Conventions of 1899 and 1907 had proposed rather indefinite rules which, in any case, were well and truly shattered by 1918. An attempt was made to produce a limitation on chemical weapons at the Washington Naval Limitations Conference in 1921–2, but this failed when France did not ratify the resolution. A League of Nations Conference in 1925 at Geneva was charged with advising the League Council on regulations to limit the international traffic in arms and munitions. No agreement on chemical weapons was reached in the framework of this conference, but an associated protocol was agreed to and initially (by 1936) ratified by 40 nations, with the USA and Japan as important exceptions. The USA ratified the protocol in 1975. The Geneva Protocol or Convention of 1925 banned the '. . . use in war of asphyxiating, poisonous or other gases, and of all analogous liquids, materials or devices . . .' together with 'the use of bacteriological methods of warfare . . .'. Further attempts during subsequent disarmament conferences to strengthen the bans on chemical warfare failed, so that the Geneva Protocol is still the only international agreement.

The weaknesses of the Geneva Protocol are apparent. Most nations ratified it with the reservations that it would only be binding among nations that had also signed, and that it would cease to bind if the prohibition was broken by an enemy state. It refers to use of chemicals, and is not a disarmament treaty. Technically, the definition of 'poisonous' leads to many problems, as detailed in Chapter 4. Tear gases have always presented difficulties with respect to interpretation of the Protocol.

The Protocol did not help to dispel the general fear that chemicals would be used in any new major war. Mustard gas was used by the Italians in Abyssinia in 1936 and by the Japanese in China prior to 1941. The growth in international tension following the rise of Fascism in Italy and Nazism in Germany led to serious consideration of defence against chemicals in Europe. This was rightly recognised as largely a problem of defence of the civil population. In the First World War aerial bombing of civilians had occurred; England suffered 1414 killed and 3416 wounded. Obviously with the development of aeroplanes bombing would be more intense in the future and, equally obviously, a chemical bomb would be an ideal choice for an aggressor. The technical problems of protection were well understood and manuals were produced (e.g. *Civil Defence*, reference 3; *Personal Protection Against Gas*, reference 4). Looking back from the present time it is surprising to see the extent of preparations before 1939, and realise the general public belief that chemical weapons would be used. As a child in England during the early 1940s I had a gas mask in a cardboard box which I had to take with me everywhere. At school I had drills in putting on the mask; the box was placed on the desk before me, and I then put on the mask and was inspected by the teacher. The task of supplying the whole population of Great Britain with gas masks was completed in 1940 and must have been an enormous effort in manufacture and logistics. Other European countries made similar efforts.

As it happened, toxic chemicals were not used in the Second World War, although all combatants had stocks ready, and a lot of scientific research and development was performed on offensive and defensive aspects of chemical warfare. The Germans had chemical stocks deployed in northwestern Europe, British, American and Soviet mustard gas was available in quantity and the Japanese had gas shells in the Pacific Islands (some were captured in New Guinea). Training in chemical defence was required of all military and many civilian populations. There must have been many occasions where a decision to use chemical weapons was near to being made.

The position was complicated by the development during the war of the 'nerve gases', organophosphorus compounds which inhibit the enzyme cholinesterase in the nervous system, causing a breakdown in nervous control of the body. One toxic compound of this family (tetraethylpyrophosphate) had been synthesised by Philippe de Clermont in 1854, who reported on its taste (*d'une saveur brulante*) but despite this tasting lived on for 65 years to die at age 90. Similar compounds were made at various subsequent times, and in 1932 the German chemists Lange and Krueger noticed the toxic effect of the vapours on themselves. The investigation of this class of chemical as potential insecticides was begun by I. G. Farbenindustrie, and Gerhard Schrader patented the general formula for such insecticides in 1937. The German military were aware of this work. A factory built at Dühernfurt near Breslau produced 10 000 tons of Tabun (GA), 600 tons of Sarin (GB) and small quantities of Soman (GD). At the end of the war this factory and its stocks fell into Russian hands. All three compounds are nerve gases of high human toxicity, a subfamily of organophosphorus compounds distinct from the insecticides.

The Allies had followed up the early reports in the German chemical literature and also began to make organophosphorus compounds. The compound attracting most interest was DFP (diisopropylfluorophosphonate), which is quite a bit less toxic to humans than GB or GD. This compound was made by a team which included Dr. B. C. Saunders, (5), who was later to become one of the more popular lecturers in organic chemistry at Cambridge. I was entertained during his lectures in the 1950s with stories of his wartime work. The toxicity of new compounds was tested on the chemists themselves. Saunders had a story of how one compound had little effect, so that the chemist relaxed in his chair in the test chamber and told his colleagues to increase the vapour concentration. The experiment was abruptly terminated when someone noticed that the subject's pupils had shrunk to pinpricks, a sign of nerve gas poisoning. There were few wartime scientists in the chemical warfare field who did not acquire some visible reminder in the form of Lewisite or mustard burns, or some vivid memory of an accident or near-accident. The Germans seemed to believe that the Allies were more advanced in nerve gas development than they actually were, a misapprehension that the Allies were unlikely to dispel.

Gas was used in the Nazi extermination camps to kill Jews and others who did not subscribe to the standards of the Herrenvolk. Zyklon B was a mixture of hydrogen cyanide with cyanogen chloride and other irritant chemicals, which were added to make the peacetime use of the insecticide safer, i.e. detectable at low concentrations. The use of this gas was not chemical warfare in the accepted sense, but was certainly a remarkably horrific example of intraspecific chemical warfare within my introductory definition,

Given all this wartime activity, why were chemical weapons not used in battle? Hardly because of moral scruples; there must have been more mundane appraisals of the situation. One answer may be in the defensive preparations that all sides had made. Chemicals are peculiar in that defence against them is relatively easy. A respirator, gas cape, and rubber boots and gloves will protect a person well whereas it is much harder to protect against high explosives or shrapnel. Therefore effective use of chemicals depends on a surprise attack. If all combatants are prepared, then chemical attack has much less value. The civil defence preparations in the late 1930s were probably a major factor in averting chemical war. One lesson from the war was that a trust in neutrality was meaningless (witness Belgium, Holland, Denmark and Norway in 1940) and that strength was the only factor respected by the Axis powers. Defence against chemicals was one of the few fields in which the Allies were prepared, and this may have saved them from chemical attack.

A second factor may have been the fear of reprisals, particularly if the Germans believed the Allies had developed effective nerve gases. The

Germans may have seen no clear advantage in the first use of chemical weapons, and the British were nervous about attacks on their cities despite the civil defence precautions. Towards the end of the war the development of the atomic bomb provided the Allies with a possible alternative to gas, an option that was eventually used at Hiroshima and Nagasaki.

For whatever reason, the world was saved from a chemical dousing in the 1940s. After the war, the military machine was largely run down. The major nations made stocks of nerve agents and disposed of some older munitions by various means. Research into the effects of nerve agents and means of defence against them became a matter of priority in defence research establishments. This was paralleled by synthetic programmes and research work in the chemical companies, universities and hospitals on insecticidal organophosphorus compounds, their toxicity and the treatment of poisoning – accidental or suicidal. We then move into the situation that exists at present, so that next we will discuss modern chemical warfare agents and their possible use.

The chemical agents still believed to be stocked by some nations are shown in Table 10.1. They include nerve agents of the G and V series, and vesicant agents such as sulphur mustard and Lewisite. These chemicals are not gases but volatile liquids. The V agents in fact are hardly volatile at all, so that they enter the body by liquid contact to the skin or by inhalation as a liquid aerosol. The more volatile compounds such as GB are principally a

Table 10.1. *Toxicity data on some obsolete and current chemical warfare agents*

Compound	Boiling point[a] (°C)	Inhalation toxicity LCt_{50} ($mg/m^3/min$)	Percutaneous toxicity (adult man)[bc]
Chlorine	−34	60 000	—
Phosgene	+8	3 000	—
Hydrogen cyanide	+26	5 000	—
Sulphur mustard	+217	1 500	E = 20 μg
Lewisite	+190	1 500	E = 20 μg
GB	+158	100	L = 2 g
VX	+298	100	L = 3 mg

[a] Mustard and VX decompose before reaching their theoretical boiling point.
[b] No entry under percutaneous toxicity indicates that the percutaneous toxicity of the compound has no practical importance; contact with liquid hydrogen cyanide is very dangerous but would not occur in warfare.
[c] E, effective dose to produce a blister; L, lethal dose. Note the change of units in these entries.

threat as a vapour by inhalation. Once the nerve agents enter the body, their action does not differ. Nervous function and control is disrupted, and death usually occurs by asphyxiation caused by the failure of the rib muscles and diaphragm to ventilate the chest. The victim thus dies of lack of oxygen. However, although this is the immediate cause of death, the toxic effects are general throughout the nervous and muscular system and are not confined to the respiratory apparatus.

Vesicant effects have been described in this chapter. Lewisite has an immediate effect, whereas the skin damage due to sulphur mustard only appears some hours after contact. Hydrogen cyanide and phosgene are liquids boiling at temperatures near ambient, so that they are effective in the vapour form. Their toxicities are about 50 times less than those of nerve gases (inhalation toxicity LCt_{50} is about 5000 $mg/m^3/min$ for phosgene and hydrogen cyanide, 100 $mg/m^3/min$ for GB), and it is difficult to know if the military regard them as potential chemical weapons nowadays. There are many more compounds of various types that are considered as chemical weapons but the main candidates are the nerve agents and sulphur mustard. These are undoubtedly stocked by the Soviet Union and the USA, and probably by some other nations (now including Iraq). What is difficult to find out is what proportion is stocked actually in chemical weapons. The US stocks of nerve agent at Tooele Arsenal in Utah are bulk stocks in drums, or in obsolescent munitions, not suitable for immediate use. No country is prepared to declare what are its stocks of chemical munitions ready for deployment and use.

The great difference between a chemical war that might happen now and the First World War would lie not so much in the nature of the chemicals used (which, with the exception of the nerve gases, would be much the same) but in the means of delivery of these chemicals on to the target. Chemicals were used rather crudely in the First War and the means of delivery were becoming better understood only as the war finished. Since 1918 there has been much study of ways to deliver the chemical on target in the most effective way, e.g. to spread liquid droplets of the right size over the largest area, or to create very high concentrations of vapour suddenly on a limited target. Any large missile can be used as a carrier of chemicals, so that a rocket can explode at an altitude of 1000 m to spread droplets of viscous chemical over several square kilometres. Public preoccupation with the dramatic effects of nerve agents has tended to obscure this great change in delivery methods. The attacker is not now dependent on the direction of the wind to anything like the degree the Germans and British were in 1915.

Biological warfare to my mind means the offensive use of living organisms which are capable of reproducing themselves, as in the classic infectious diseases such as cholera, smallpox, etc. However, some people also include toxins derived from living organisms as being tools of biological warfare, although these toxins cannot reproduce themselves and are capable of being synthesised chemically, in principle at least. Therefore I cannot accept the view that toxins are distinct from chemicals, and believe that the use of biotoxins in war constitutes chemical warfare. Such use is of course not new since most of the early known poisons were extracted from living organisms (curare, belladonna, etc.). It has, however, become of topical importance with the recent accusations from the USA that the Vietnamese have used biotoxins from naturally occurring moulds to attack insurgents in Laos and Kampuchea. The further claim has been made (6) that the Soviet Union has experimented for years with mycotoxins and supplied them for use in the Yemen as well as southeast Asia. There are two separate questions resulting from the recent reports, firstly have chemicals been used and secondly were the chemicals mycotoxins? The answer to the first question is that use is not proven but credible. To the second question the evidence to date (1985) is not conclusive, and there are serious doubts about the effectiveness of the mycotoxins. If the aim were to kill, then GB or VX would be much more effective. If the aim were to harass and irritate, other chemicals are available which harass without killing, but the reports speak of many deaths. Seagrave's main thesis in his book is that the Soviets have a new generation of biotoxin weapons which are much more effective than nerve agents. What little quantitative data he gives indicate toxicities of about the same order of that of the nerve agents. There is no doubt that very highly toxic biotoxins exist; what I doubt is that the Soviets have them in munitions which they are willing to give to rather untrustworthy allies.

Most of the specimens of 'Yellow Rain' collected in Laos or Kampuchea have proved to be the droppings of bees, coloured yellow with pollen. The vast majority of samples contain no toxin. What seems to have happened is that people have been sent back to an area of alleged chemical attack to collect samples of 'Yellow Rain'. The only yellow material they could find was the bee product, so that is what they brought back. The intense concentration of effort on these samples, which are irrelevant to the chemical warfare issue, means that we will probably never know now which, if any, chemicals were actually used. The scientific evidence about 'Yellow Rain' is well summarised in reference 7.

Evidence for the occurrence of chemical warfare in the Gulf War is much more definite. The team of international observers that went to Iran in March 1984 at the request of the United Nations Secretary General was able to sample chemical bombs (Fig. 10.1) and to inspect the victims (Fig.

Fig. 10.1 Iranian Revolutionary Guards sample a chemical bomb. The bomb failed to detonate and was recovered from marshy ground, from which it was transferred to this stony area. (*a*) The bomb is being tilted to allow more mustard 'gas' to flow from the nose, through the hole left by the removal of the fuse. A closeup in (*b*) shows the sample bottle being filled. Hoor-Ul-Howaizeh, March 1984 (photographs: Dr P. Dunn).

(*a*)

(*b*)

10.2) of attacks with mustard and nerve gas (8). The evidence was too extensive to have been the result of an Iranian fabrication, and it points very clearly to the use of chemicals by Iraq. This occasion was the first time that nerve gas (GA) is known to have been used in war.

At the present time attempts are still being made to secure a chemical disarmament treaty. The United Nations is sponsoring the Conference on Chemical Disarmament which meets regularly in Geneva, and the USA and USSR have had bilateral talks. The difficulties are immense, and I will outline a few of the problems.

The initial difficulty is to define a chemical weapon. An artillery shell full of GB is obviously one, but what of a cylinder of phosgene in a storage depot, or a drum full of an organophosphorus compound which is the intermediate in the synthesis of *either* nerve agent *or* insecticide? The modern chemical industries produce quantities of substances which are potentially useful in a chemical weapon (e.g. methylisocyanate, which was an effective killer in Bhopal). One approach is to distinguish these on the basis of human toxicity. Thus supertoxic chemicals (see Table 4.2 and Chapter 4) would be selected for examination and the nation possessing them would have to justify their presence on the basis of a clear, non-military use. These twin criteria of toxicity and intended use would not define the cylinder of phosgene mentioned above as a chemical weapon, yet such cylinders were very effective weapons in the First World War.

The second problem centres on the problems of verifying a chemical disarmament treaty. It is extremely unlikely that all the nations involved will consent to completely free inspections of all industrial and military installations, therefore some more remote method of monitoring the treaty has to be found. One idea is that the raw materials for chemical agent synthesis should be monitored, so that if a country starts using more organophosphate intermediates than its production of insecticides warrants, then it is challenged and required to explain the use. However, it is unlikely that such balance sheets of usage and production would be accurate enough to show up a steady, moderate but important production of nerve agent. Another idea is that very sensitive, modern analytical techniques be used to monitor the wastes from chemical factories to see if traces of chemical agents or related compounds are present. Many (but not all) nerve agents contain a bond between a phosphorus atom and the carbon atom of a methyl group. Such a bond is stable and occurs in very few natural compounds. Therefore its presence in substances isolated from river water would indicate a questionable activity in any upstream chemical factory. Its absence would mean nothing; GA, which does not have this bond, could be in production, or all could be innocent and in order.

Fig. 10.2 The victims of chemical war. The men in (*a*) are survivors of an attack with nerve agent (GA or Tabun) and are shown in the Tafti Stadium Infirmary, Ahvaz. The corpse in (*b*) was the victim of mustard gas poisoning. Coroner's Mortuary, Teheran, March 1984 (photographs: Dr P. Dunn).

The difficulties in chemical disarmament therefore reduce down to trying to spot unusual chemicals among the enormous industrial production of them, and to identify chemical-carrying missiles among the great numbers of munitions held by all nations. It would be much easier to conduct chemical disarmament as part of a general disarmament process. Why then the attempt to treat chemicals separately? Some may argue that this is because the use of chemicals is a particularly horrible form of warfare. It is difficult to accept this, for the mutilation and agony caused by machine-gunning, shrapnel, blast and burning of acceptable 'conventional' warfare cannot be any less horrible than the effects of chemicals. Probably the main reason to seek chemical disarmament is that chemical war has not occurred to any extent since 1918, that it is a form of warfare which is distinct from the conventional, and therefore its banning will limit the variety of horrors to which man may expose himself, if not eliminating such horrors altogether. The use of chemicals in the Gulf War has further prejudiced the chances of a treaty.

In summary, we have used chemicals on our own species for one intensive period of three and a half years, have escaped so far the consequences of further, large-scale chemical war, and have a slight hope of banning further such wars if some very complex political and technical problems can be solved.

References and further reading

1. Prentiss, A. M. (1937). *Chemicals in War.* 1st edn. New York: McGraw-Hill.
2. Foulkes, C. H. (1934). *'Gas': The Story of the Special Brigade.* Edinburgh: Blackwood.
3. Glover, C. W. (1938). *Civil Defence.* London: Chapman & Hall.
4. Anonymous (1937). *Air Raid Precautions Handbook No. 1, Personal Protection Against Gas.* London: H.M.S.O.
5. Chapman, N. B. (1984). Obituary Notice. Bernard Charles Saunders 1903–1983. *Chemistry in Britain*, October 1984, p. 917.
6. Seagrave, S. (1981). *Yellow Rain.* 1st edn. New York: Evans.
7. Ember, L. R. (1984). Yellow Rain. *Chemical & Engineering News*, Jan. 9, 1984, pp. 8–34.
8. Andersson, G., Dominquez, M., Dunn, P. & Imobersteg, U. (1984). *Report of the Specialists appointed by the Secretary-General to investigate Allegations by the Islamic Republic of Iran concerning the use of Chemical Weapons.* UN Security Council document S/16433, 26 March 1984.

For further reading, the SIPRI volumes (A) give a lot of information on all aspects of chemical warfare and disarmament. Reference B is a reasonably accurate history of chemical warfare, perhaps inclined to sensationalism. The last reference (C) needs to be read fairly critically, for it tends towards sweeping statements and inaccuracies in detail; it presents a point of view worth reading given these reservations.

A. Various authors: *The Problem of Chemical and Biological Warfare.* Six volumes produced by SIPRI (Stockholm International Peace Research Institute) in early 1970s. Stockholm: Almquist & Wiksell.

B. Harris, R. & Paxman, J. (1982). *A Higher Form of Killing. The Secret Story of Gas and Germ Warfare.* London: Chatto & Windus.

C. Murphy, S., Hay, A. & Rose, S. (1984). *No Fire, No Thunder.* London: Pluto Press.

11

Synthetic chemicals in the environment: pollution or assimilation?

I find that I have planned to discuss the problems of environmental pollution in one chapter. Clearly no one can cover this topic in detail in such compressed form, therefore I will discuss the concepts in outline, and leave you to follow up the detail. The technical aspects are briefly but clearly described in reference 1, and more extensively in 2. The first question is: what is the environment? It is everything around us, living and inanimate, synthetic and natural. None of the components of the environment is inert and unaffected by the others. The barest sandstone rock was once the home of marine creatures, is now a shelter for lichens and insects, and will become once again a sandy bed for plant and animal life to thrive in. When pondering the unity of the environment, we can draw on the imagery of the former dean of St Paul's, concerning for whom the bell tolls. To paraphrase Donne, no part of the environment is insulated from the continent or main, any loss is a loss to the whole. Leaching of soil from a hillside, the loss of a minor aquatic animal, or the destruction of nesting sites of a bird species, all these actions have an effect on the total environment. In our eyes this effect may seem important or trivial, but the bell tolls for us as much as for any other animal or plant species. The environment is a net reaching to us all; only the wilfully blind can ignore the web.

We, of course, see the environment from our self-centred human viewpoint. What we see is an environment very largely controlled by us and altered very greatly from its pristine state. This preoccupation with our own society should not blind us to the presence of many other societies or populations of organisms, which flourish more or less obtrusively alongside our human society. Insects abound in great numbers, but we do not notice them until they annoy us. Most do not. Thus I am willing to bet that the mass of the ant population in Australia greatly exceeds that of the human, but we only notice them when they invade our homes. Similarly, the microorganisms of soil live in complex interrelationships with one another and other creatures, yet we are not even aware of them until we start to look at the breakdown of humus, or at the decomposition of chemicals in soil.

The environment that concerns us humans can be regarded as of varying scope, depending on the problem we are considering. It may be very limited if we are considering a specialised workplace, or very wide if we want to study the population of a country. Here I will generally be writing about the wider scope, and ignoring specialised industrial problems.

The next question is: what concepts can best express the social problem of synthetic chemicals in the environment? After some thought on this, I believe the problem is essentially one of the economics of distribution and collection. Raw materials are collected in bulk, converted into synthetic chemicals, fashioned into articles or formulated for use, then distributed to the consumer. This is economically favourable to the manufacturers and distributors; they profit from the price the consumer pays. The products are now scattered around towns or countryside, and the wastes from manufacture are also distributed in the environment. Ideally the residues should be collected and destroyed, but of course this is an uneconomic process from which no profits can be made. If it is to be done, and pollution to be reduced, society has to pay. The problem is obviously lessened if the synthetic chemicals can be assimilated in the environment, and to enable this to occur is a technical challenge that is now being addressed. However, the basic problem is one of economics, as in most of our social problems. Note that I am not making a political attack on private enterprise, for effective legislation is the only way to direct and control both public and private organisations (see Chapter 15). Therefore the social problem of environmental pollution has the opposing elements of easy and profitable distribution countering the difficult and uneconomic collection or destruction of residues and wastes.

The increasing amounts of synthetic chemicals that we are making (Chapter 2) are being returned to the environment after use. I use the word 'returned' because the chemicals had to come from the environment anyhow. What we have done is to alter the chemical nature of these materials quite drastically, and also redistribute them. Thus crude oil and gas from Saudi Arabia is transformed into plastic containers, which are then liberally scattered over North America, Europe, Australasia and practically everywhere else. Or lead is mined from Broken Hill or Missouri, combined with four ethyl groups and distributed in petrol, to be aerosolised into the air of every town and city in the world (300 000 tonnes a year in the USA). Therefore the disposal of chemicals involves the study of chemical change and distribution. Disposal of course does not necessarily mean pollution. If the material is taken out of a hole in the ground in Saudi Arabia and eventually returned as a polyethylene container to a hole in the ground in Melbourne, then the process is unobjectionable. It is aesthetically objectionable if the container is left lying around in a city park (Table 11.1). Ideally, we would like what we have taken from the environment to be returned to it in an assimilable form. If we persist in the use of synthetic chemicals, then disposal is inevitable. The question is therefore can this material be returned to the environment and assimilated, or is pollution inescapable? The answer must be sought through chemistry and distribution.

Pollution to date has resulted from at least three causes: the use of fuels to provide energy; the metals mining and fabricating industry and, finally and most recently, the New Chemical Age. I shall try and talk mainly about the latter, but the most dramatic effects so far have been from pollution due to the two former activities. Thus the hills around Queenstown in Tasmania are bare of trees or much other vegetation, although areas further away from the town are well wooded. This denudation was due to the felling of the trees to act as fuel for copper smelters, the sulphurous fumes from which killed off other vegetation. Air pollution is very obvious; I mentioned London fogs in Chapter 8, and in recent times photochemical

Table 11.1. *The types of objectionable environmental effects produced by the four main classes of synthetic chemicals (an anthropocentric viewpoint)*

Class of chemical	Form as present in environment	Type of objectionable effect[a]
Structural	Plastic containers, polythene bags, fishing line	Aesthetic, possibly economic
Pesticides	Parent chemicals, residues, used containers	Health, economic
Drugs	Environmental quantities probably negligible	Health
Process chemicals	Parent chemicals and breakdown products (stable chemicals greatest problem)	Health, economic

[a] Aesthetic effects are mainly visual; economic effects embrace damage to useful animals and plants; health refers to factors deleterious to humans.

smog has become synonymous with Los Angeles. Pollution by synthetic chemicals has been more subtle in effect. A large part of the literature on pollution refers to mining wastes, heavy metals in the environment, and pollution from energy production, and is therefore not strictly related to my discussion. Firstly I will describe the environment further, then the chemical sources of pollution and, finally, the prospects for future control.

The environment is made up of non-living matter which has been colonised everywhere by living organisms of varying degrees of structural complexity and chemical ingenuity. I suppose the crater of an active volcano is free of life, but there cannot be many other places on the earth's crust that are sterile. We must not think that the non-living part of the environment is inert. It is subject to constant physical fluctuations, as in heat, light and other radiation, which produce wind, water movement, cracking of soils and erosion, drying of some areas of land and flooding of others. The air is always in motion, its movements directed by complex heating and cooling patterns and the motion of the earth. Similarly, water is always in motion, particularly on land and the surface of the sea. Earth and water are closely associated, so that the surface of the earth is being continually modified by water movements. Earth itself is chemically active, which has significance for chemical pollution. Clay soils have the capacity to absorb many chemicals, so that they may be retained near the point of contamination. Sandy soils have little capacity for absorption and chemicals are carried rapidly through these in ground water, if they are soluble in water.

This already complex system has been colonised by living organisms to form a system of extreme complexity. The organisms live in relationship to the physical environment and to one another. Thus they are interdependent for food and shelter, and the food chains may be long with many stages from microscopic creatures to large predators.

I want once again to emphasise the interdependence of all elements of this enormously varied living and non-living world. Give it a kick in one place and you may put the whole lot out of gear. Pour in a synthetic chemical in one river, and the result will be unpredictable.

Humans are part of this system, but as our population has increased and become concentrated in towns, our natural wastes have burdened the environment. Mohenjo-Daro was equipped with sewers 4000 years ago, but it was not until the mid-nineteenth century that sewerage works of the modern type were designed. These had to cope with growing quantities of human waste and also to contend with industrial pollution from the metal industries. The environment in general also had to contend with air pollution from smelters and coal fires, the water was contaminated with wastes from metal and paper industries and from inorganic chemicals leached from mine spoil dumps. This pollution created most of the environmental damage we see today. When the New Chemical Age arrived 40 years ago, the wastes from the production and use of synthetic chemicals further burdened a system of controlled disposal nearly at its limit, and threatened to add to the problems of uncontrolled disposal. Having done a spectacular job of environmental destruction over the last 100 years, we are reluctant to repeat the performance with synthetic chemicals.

Parallel to the disposal of chemical waste through the urban disposal systems is the direct release of chemicals to the environment. The major example of this is the use of pesticides, but chemicals are also discharged directly to the air from the factory and the home.

The routes of release of chemicals to the environment are shown in Fig. 11.1. The chemical factory may release chemicals directly to the air, or the discharge may have to be cleaned first, depending on local regulations. The same applies to liquid waste. It may be acceptable in the local sewer, or it may have to be treated first. Discharge directly to rivers may be possible, but this would have to be very carefully monitored. These methods of disposal are available for water soluble wastes, or those that can be made so. If the chemical waste is an oily liquid, the alternatives are to incinerate it, or

Fig. 11.1 Some routes for the release of chemicals to the environment.

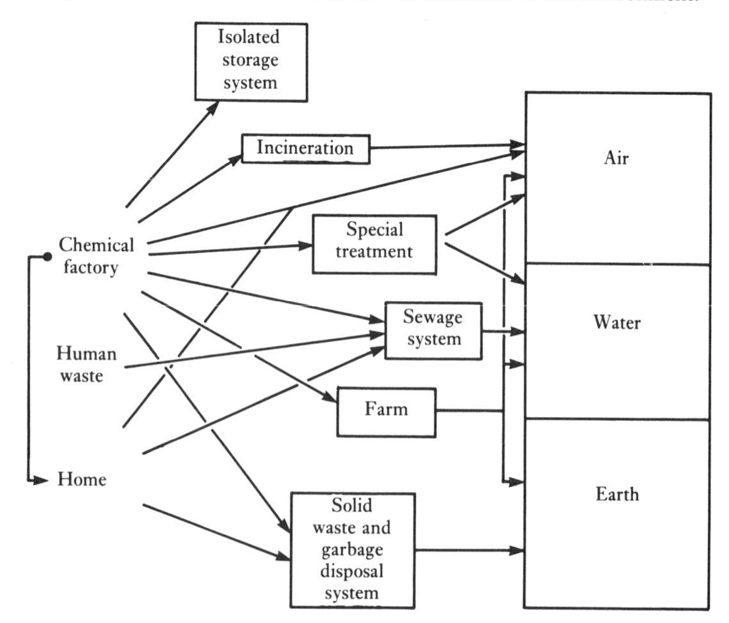

to dump it. Neither alternative is attractive. Incomplete incineration can result in the production of carcinogenic chemicals, in a manner similar to barbecueing a steak (Chapter 8). The fly ash from municipal incinerators has been found to contain polychlorinated dioxins and furans, both very toxic classes of compounds. There is some dispute as to whether these chemicals arise from the incineration of closely related precursors, or more generally from many chemicals (3). Anyway, to be effective, incineration has to be done at high temperatures with provision for cleaning the exhaust gases. This is extremely expensive, both in capital costs for the facility and in running costs. The alternative is to dump the liquid waste. Then the problem is to find an isolated system that will keep the liquid away from the general environment. The idea is to dig a pit in an impervious soil in a geologically stable area. If such an area can exist in theory, it is certainly not easy to find. The disposal of liquid oily waste is thus the great industrial problem, which we will discuss further. Solid factory waste which is truly solid and not very toxic can go in the municipal tip (if the authorities are cooperative). The worry then is whether anything can be leached out of the solid by ground water and appear elsewhere in the environment.

Chemicals released from the home cannot be neglected. Direct discharge to air has only been questioned in one case as far as I know, that is the matter of propellant release from aerosol cans. The propellant vapours have concentrated in the upper atmosphere sufficiently to interfere with energy transfer and reduce the ozone concentration (4), thus increasing the 'greenhouse effect'. No one seems to have determined what use is made of backyard incinerators to destroy plastics, and whether any hazard results. Most chemicals in domestic use are probably finally disposed of by flushing down the drain (if liquids) or taken to refuse tips (if solids).

Farm chemicals are usually dispersed over the countryside, and their subsequent fate then depends on their physical and chemical properties. They can end up in air, water or bound to earth particles.

Following this brief account of the forms of pollution, it is necessary to outline the fate of chemicals in the environment. Firstly, it must be realised that this is a complex technical subject. Quite the wrong impression may result from a superficial study of the topic. All chemicals are different in their behaviour in the environment. If you want to know what happens to one particular chemical, you must search for information on that chemical. If no information is available, some generalisations from chemical class to class are possible, but require caution. As an example, I had to advise on the possibility of leaking phosgene contaminating the water supply to a city from a site about 100 km away. As phosgene is completely decomposed in water in 20 s, it was easy to show that no problem existed. On the other hand, some chlorinated hydrocarbon insecticides are very stable in water, although their hazard is limited by the highly insoluble nature of these compounds.

The life of a chemical in the environment is dictated by its chemical stability and its tendency to bind to other environmental components (usually soils) as shown in Fig. 11.2. Thus it may be broken down (often accelerated by microorganisms) or it may bind to the soil, and perhaps be released again if the environmental conditions change. Alternatively, the chemical may be so inert that it stays where it was dumped, or it may move directly into wide areas of the environment without change. In practice, the former condition never applies absolutely; some traces of the most inert chemical will be found at a distance from the point of discharge or dumping.

Toxic chemicals which are chemically reactive as a consequence usually show both an acute form of toxicity, and a short life in the environment. Conversely, those which are unreactive have an undramatic but serious chronic toxicity and persist in the environment. This is illustrated in Table 11.2, which gives some examples of chemicals. There is thus quite a division between those chemicals which are dramatically toxic but not a long-term pollution problem, and those which are stable and thus are chronic pollutants. We cannot of course ignore the first group; methyliso-cyanate was a disastrous pollutant for a short while at Bhopal. These chemicals are problems in terms of handling safety; the latter group are truly an environmental problem.

Fig. 11.2 Possible fates of a chemical in the environment, excluding uptake into biological systems. Fate is dependent on the physical and chemical properties of the particular chemical.

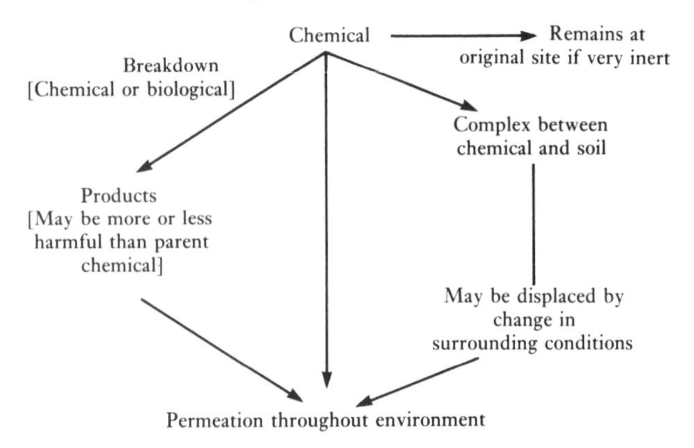

We can now discuss some examples of environmental pollutants. First we can survey the classic examples of those chemicals which are environmentally stable and chronically toxic, and owe these properties to their content of halogen atoms (usually chlorine). Table 11.3 lists some of these compounds in functional groups. The insecticides of this type pioneered the control of insects by synthetic chemicals after 1945 but, by the late 1950s, there were doubts about them which resulted from their appearance in a wide range of animal, plant and soil samples. The then new technique of gas chromatography enabled these compounds and their

Table 11.2. *The consequences of chemical stability as a determinant of the pollution potential of chemicals*

Stability (Lifetime in environment and in body tissues)	Consequences	Examples
Unstable (Minutes to hours)	Acute toxicity Short life in environment	Organophosphorus and carbamate insecticides; phosgene, methyl isocyanate
Stable (Months to tens of years)	Chronic toxicity Long life in environment	PCBs, PAHs, Dioxin, Chlorinated hydrocarbon insecticides

Table 11.3. *Examples of stable halogenated hydrocarbon chemicals that have caused pollution problems*

Functional class	Examples	Current status
Insecticides	DDT, BHC, Aldrin, Dieldrin	Many countries ban widespread use in agriculture; still used for special insect control problems
Herbicides (contaminant of)	Dioxin	Concentration in herbicides is controlled in many countries (to 0.1 ppm)
Electrical insulators	PCBs	Use discontinued in last few years; still much electrical equipment in service which contains PCBs
Fire Retardants	PBBs	Similar situation to that for PCBs
Refrigerants and propellants	Chlorofluorocarbons (Freons, etc)	Still used as refrigerant fluids; use in aerosol cans discouraged

residues to be found for, without the increased sensitivity of this technique over older analytical methods, we would have been ignorant about the spread of chlorinated hydrocarbons throughout the environment. This is another case in which analytical capability has dictated the degree of appreciation of a problem (Chapter 6). By the mid-1960s the use of these insecticides was declining due to restrictions imposed by legislation. However, they are still to be found in biological and environmental samples.

PCBs and PBBs (polychlorinated biphenyls, polybrominated biphenyls) were selected for use because of their great stability. Thus PCBs are, or were, used as the insulating medium in electrical transformers and big capacitors. They are liquid or semi-solid. A capacitor may be constructed with layers of paper soaked in waxy PCB wound around the plates. If the capacitor fuses and ruptures the outer envelope, PCB is exposed to the outside, to the maintenance person and to the environment. The health effects of PCBs at low levels are not dramatic in the short term, they are believed to harm the liver and probably other organs after long-term exposure, but the main concern is related to their long life in the body once they enter (through the skin). The toxic effects of some batches of PCBs may be due to contaminants such as dibenzofurans, as in the example of 'Yusho' disease caused by contamination of rice oil by PCBs in Japan in 1968. PBBs are in general similar in properties to PCBs. Extensive pollution by these compounds occurred when PBBs were inadvertently mixed with cattle feed in Michigan in 1973. By the time the resulting animal sickness had been traced to its cause, the contamination had spread via food products to the general population. Most of the 9 million inhabitants of Michigan subsequently had detectable levels of PBBs in their body tissues. Whether or not this has led to any definite increase in disease in that state still seems to be a matter of debate. Whatever the result, the incident illustrates the contamination that is possible from one single polluting accident with a stable chemical.

A related problem has occurred with dioxin. Dioxin appears as a toxic threat in two situations; as a contaminant in 2,4,5-T herbicide and as a by-product in waste from the manufacture of trichlorophenol and chemicals further elaborated from that phenol (e.g. hexachlorophene disinfectant and 2,4,5-T). The first situation is considered in Chapters 13 and 14, where it is seen that it is debatable whether dioxin in 2,4,5-T has had any observable harmful effect. The second situation, which is considered here, is an example of gross pollution with a very toxic chemical. The important question is whether the two situations are equivalent or whether they represent two different phenomena that ought not to be related.

One example of waste oil contamination with dioxin is the history of events that led to the Environmental Protection Agency (EPA) recommendation that the Federal Government buy out the small town of Times Beach in eastern Missouri, because of persistent dioxin contamination. The story started at a plant in Verona, Missouri, that had produced 2,4,5-T and hexachlorophene. Verona lies across the state from Times Beach, in the southwestern corner. In 1971 the company that then controlled the plant contracted out the removal of waste bottoms or sludge from the chemical plant. This oily waste (70 000 l) was carried across the state by a waste haulier to a store at Frontenac, a suburb of St Louis. The contractor used some of this oil to spray on horse arenas in order to settle and compact dust and loose earth into a hard surface. This activity started in May 1971 and resulted in the sickness and death of many animals, including death of 65 horses. Four children and one adult showed symptoms of dioxin poisoning but recovered. There was of course nothing to link this strange outbreak of disease with dioxin. The waste oil continued to be used as a spray on many sites (perhaps up to 150) in eastern Missouri until 1974. In that year the investigations into the original 1971 poisonings finally became productive when the Center for Disease Control found dioxin at 33 ppm in the arena soil and thus were able to link the disease with the source of the waste oil. This led to the discovery of a storage tank back at the factory in Verona which contained waste sludge contaminated with 356 ppm of dioxin. This was detoxified at the current owner's expense.

From 1975 the issue remained dormant, until in 1979 some dioxin wastes were found buried on a farm near Verona. Investigation of this dump made the EPA slowly realise that dioxin was not decomposing in soil or on the contaminated ground surfaces, and therefore it became obvious they should survey the areas previously known to have been contaminated. Results published in 1982 showed that dioxin was present at the dump sites and horse arenas up to concentrations of 2 ppm.

The township of Times Beach stands on Interstate Highway 44 about 36 km southwest of central St Louis. Here dioxin was found at 300 ppb, presumably having been redistributed from the store at Frontenac. Fortunately this contamination remained in place after flooding of the nearby Meramec River in December 1982. In February 1983 the EPA offered to buy contaminated properties in Times Beach, but the actual movement of people and payment of compensation was stalled by the usual problem of which authority was actually responsible for assuming title to the land, paying compensation, and cleaning up afterwards.

Times Beach is not, of course, the only contaminated area. If we reconstruct the fate of dioxin-contaminated sludge from the Verona factory, there are four main routes of disposal:

1) disposal by approved means before 1971;
2) removed by contract haulier to Frontenac in 1971, not in an approved manner but probably not illegally (subsequently sprayed on many sites in eastern Missouri);
3) stored at Verona and destroyed in 1974–5;
4) illegally buried in dump at Verona, and found in 1979.

Item 2 presents the most problems. The complete data for dioxin concentrations at all possible sites sprayed are not yet available. The actual testing program is costly, let alone the costs of compensation, moving people and cleaning up.

This Missouri episode (for which I have drawn on reference 5 for details) is not the only occurrence of dioxin contamination. The Love Canal (New York State) situation and the various factory explosions which have released dioxin are described by Hay (3). Of the latter, Seveso was the most publicised factory accident. The initial great concern about Seveso stemmed from our uncertainty about the risks to the population living around the Seveso factory. In retrospect, the long-term health effects seem less than was feared, but of course the elapsed time is still too short to enable us to make the final judgement.

The 1970s were thus bad years for the manufacturers of halogenated chemicals. The United States finished the decade with dump sites and abandoned factory sites all over the country, contaminated to a greater or lesser degree (usually unknown) with PCBs, PBBs, dioxin and other intractable chemicals. Dump sites are a miscellany of unrecorded drums and containers (Fig. 11.3) presenting unknown hazards to people and to the environment. A clear, popular account of the present situation is given in reference 6. At October 1984 there were, in the USA, 786 waste sites designated or proposed for listing on the National Priorities List of the EPA. It is estimated the total could rise to 2500 sites. Meanwhile, the 'Superfund' set up by Congress in 1980 has been exhausted, and Congress is expected to have to provide much more funding than the initial $1.6 billion. The situation persists due to lack of money and of officials to enforce the regulations.

It should be pointed out that these dumps contain many chemical wastes other than chlorinated hydrocarbons. The latter are the most intractable problem, however. The only way of dealing with them is to store them perpetually in as secure a containment as possible. This leads to engineering and logistical problems, for we are not talking about small volumes of pure chemicals, but vast amounts of contaminated soil and water. The concentration of actual chemical may be quite low, perhaps 100 ppm, or 10 ppm, or less. D'Appolonia (7) describes the clean up of an

880 acre site surrounding a hexachlorocyclopentadiene plant (the chemical is a precursor of the insecticides aldrin and dieldrin). This involved moving 800 000 cubic yards of contaminated earth into a vault lined and sealed with 10 foot thick clay walls. This kind of operation in the USA and other major chemical-producing countries will result in large areas of land being denied for other use. We will have a perpetual legacy of storage sites which will require regular monitoring and maintenance to ensure that nothing leaks out. This is one aspect of waste management. Many other technical aspects, and the relevant legislation (in the UK), are discussed in reference 8. We should be concerned about temporary storage depots for chemicals, as well as waste dump sites, for accidents such as fires or explosions will release chemicals to the environment (Fig. 11.4).

A very different example of environmental pollution is that of Minamata disease. This was caused by mercury from a plastics factory, so that it was not a synthetic chemical itself that polluted Minamata Bay, but a metal used as a catalyst in production of synthetic chemicals. The disease was recognised in 1953 and related to mercury in the bay, but the real explanation was not available until 1969. The causative agent was dimethylmercury, formed from metallic mercury by bacteria in the mud at the bottom of the bay (1). Minamata disease was expressed as damage to the

Fig. 11.3 Chemical dump sites like this present an environmental threat if liquids can leak into the soil and spread to water supplies. The immediate site is a toxic threat and an aesthetic horror (photograph courtesy of Battelle Memorial Institute).

nervous systems of the local inhabitants, and also by the appearance of birth defects. The lesson is that we must anticipate problems that are, at present, technically unexpected. In 1953, metallic mercury was known to be toxic but not to the degree that would cause the disease observed. Only much later were we able to find the immediate cause of the problem. We should therefore be cautious in our predictions about the likely outcome of chemical pollution, and qualify our statements in accord with our partial knowledge.

There is a sequel to the Minamata story which illustrates how problems can keep returning in different forms. The popularity of small radios, cassette players and calculators, and the use of exposure control systems in cameras which require electrical power, has led to the development of small batteries which contain mercury. These are the button-shaped silvery batteries in your camera or calculator which you throw in the rubbish when

Fig. 11.4 The contribution of accidents to environmental problems. A fire in April 1985 at a transport depot for chemicals in Melbourne led to a potential environmental problem. Many different chemicals were released in the fire, were partially burnt and perhaps changed to new compounds. Fortunately the liquid waste and water from the firefighting hoses were dammed in a local stream to prevent further pollution of waterways. The State Environmental Protection Agency officer contemplates the problem of clean up (photograph copyright *Herald and Weekly Times*).

they are exhausted. In Japan, the motherland of electronic gadgets, it is becoming recognised that the mercury in these discarded batteries may find its way out of the garbage tips as dimethylmercury, or a similar compound, and move through the environment to water supplies (9). It is uncertain what the risk is, but we can safely bet that the usage of such batteries will increase much beyond the present level, distributing mercury into the environment near to human dwellings. Perhaps not a dramatic threat, but a development that should cause concern and should be properly evaluated.

Mercury is thus an environmental problem associated with the New Chemical Age but, of course, not a product of it. In one case it is used for production of synthetic chemicals (e.g. Minamata); the putative battery problem is a result of the Electronic Age.

These brief examples of environmental contamination are illustrative, not comprehensive. You must remember that we are very anthropocentric in our reporting and discussion of problems; what we are doing to other inhabitants of our environment is much less newsworthy than the dramatic, but not globally significant, acute incidents which usually make the headlines.

Also remember that a vast amount of pollution occurs from synthetic chemicals that are not directly toxic at all. All of our used structural chemicals, packaging, textiles, household goods, engineering plastics, all have to be got rid of, and the volume is great. Perhaps this is the greatest pollution problem of them all, how to cope with the sheer volume of relatively innocuous junk. The archaeologists of 2000 years hence will recognise city sites by the elevated tels composed of plastic debris.

This then leads to the question which I posed earlier; can we organise for our chemical rubbish to be assimilated by the environment? In the long term we probably can, but at the moment there are a lot of technical problems unanswered. We have made progress with the concept, for the bulk of insecticides used now (organophosphates and carbamates) are rapidly broken down in the environment as opposed to the earlier chlorinated hydrocarbons. The term biodegradable is popularly associated with detergents, but you should be careful not to accept this as the final answer. The detergents are biodegradable to the extent that the main nuisance (excessive foaming of the detergent effluent in waterways and sewage works) has been abolished, but other problems persist, principally now the phosphate that the new detergents contain. This stimulates excessive growth of algae in the waterways to which the treated sewage is discharged.

It is also theoretically possible to make structural plastics assimilable to the environment. Use has to be made of an environmental property that will stimulate breakdown, or we have to rely solely on the passage of time. Unfortunately for us we mostly require structural plastics to be resistant to the environment. Thus we could make use of the ultraviolet radiation in sunlight to breakdown plastics, but not if we wanted them to survive sunlight during the time of their normal function. Similarly, we might design plastics that are broken down by microorganisms in the soil, but not if the plastics were to be used for drainage pipes. Thus we can design some degree of environmental breakdown into structural materials if they are to be used indoors, but outdoors use is another problem. Building a simple time-dependent degradation into a plastic is possible for transient uses, such as packaging materials, but not for more permanent applications. No manufacturer of plastic gutters is going to even hint that his product is perishable. Some of these problems are discussed in reference 10. There, Scott surveys the chemistry of accelerated decomposition in the environment, and mentions some of the problems. The scale of use is illustrated by the usage of polyethylene sacks for agriculture in the UK, which was 100 000 tons in 1971. He also mentions the increasing use of polypropylene for ropes and fishing nets. Both polyethylene and polypropylene are stable unless purposefully modified. Porteous, in the same volume, reports on the composition of domestic refuse. Rather to my surprise, the plastic component of city waste (Birmingham in 1969) was only 1%, as opposed to paper at 48%. I think the plastic component must increase in proportion; can our forests last forever?

There is some hope that we will slowly find the technical means to make our chemical wastes more acceptable to the environment and to deal with our existing problems (Fig. 11.5). Remember, we can never expect the environment to assimilate everything without being itself disturbed or altered. Even if we reduce the material to the most innocuous form, the sheer volume of output will always be a problem. I mentioned biodegradable detergents above, but it is obvious that any chemical will have some effect where it is discharged. We cannot make it disappear, our waste is subject to the law of conservation of matter, just as any other material is. Even if we reduce it down to carbon dioxide, and the oxides of sulphur, nitrogen, etc. by a mammoth programme of incineration, we still have an air pollution problem. Once we take that crude oil out of the ground in Saudi Arabia, by that very act some degree of pollution is inevitable. Chemical technology may alleviate the situation, but will not make it go away.

To sum up this discussion on assimilation, I believe that technical progress can, and will, be made towards producing chemicals that can enter the environment and cause minimal problems. This will be a slow development. A degree of governmental pressure will be necessary to persuade industry to spend money on the research and development of new products which, at the outset, will be more costly and less competitive.

Now, let us draw what conclusions we can from this chapter. Firstly, I state again that this treatment of a complex problem is so scant that you must read much more extensively for yourselves. There are many relevant topics I have not even mentioned. One conclusion is that pollution of the environment is largely an economic problem. It is profitable to distribute goods, but costly to collect the ensuing wastes. As a consequence, government has to play the main role in controlling pollution. However, those chemical companies with a degree of foresight will be studying the technology of environmentally acceptable chemicals now. My other conclusion is that it is the slow, undramatic accumulation of chemical residues in the environment that is the main threat to our future. There is the distribution of residues by natural mechanisms throughout the living and non-living environment, and there is the impoundment of wastes by us in tips and dumps. The first waste problem threatens us indirectly through the destruction of our environment. The second problem imposes a social cost in that more and more areas become dumps which are ever afterwards

Fig. 11.5 This geophysical survey unit has been developed by the Battelle Institute, and is being used to help locate abandoned buried waste sites (photograph courtesy Battelle Memorial Institute).

useful neither to people nor to nature. At the present time there is great controversy about the siting of a liquid chemical waste dump near Melbourne. It is hard to find an area with the suitable geological features, and harder still to convince nearby residents that the operation is safe.

The conclusion therefore is that chemicals that are distributed have to be as assimilable as we can make them. Those processes which produce intractable wastes have to be changed or discontinued. In effect, the producer of chemicals will have to produce a full statement on their fate right down the line from wholesaler, retailer, user and disposal agency to the environmental scientist who checks the final assimilation. Without such action we will slowly and almost unnoticeably, but quite steadily, choke on our chemical produce. However, we do have the technical means to counter this fate, all we need is the will and the finance to use them.

References

1. Ottaway, J. H. (1980). *The Biochemistry of Pollution.* London: Edward Arnold.
2. Holdgate, M. W. (1979). *A Perspective of Environmental Pollution.* Cambridge University Press.
3. Hay, A. (1983). *The Chemical Scythe: Lessons of 2,4,5-T and Dioxin.* New York: Plenum Press.
4. Breuer, G. (1980). *Air in Danger.* Cambridge University Press.
5. Long, J. R. & Hanson, D. J. (1983). Dioxin issue focuses on three major controversies in US. *Chemical & Engineering News*, June 6, 1983, 23–35.
6. Boraiko, A. A. (1985). Storing up trouble – hazardous wastes. *National Geographic*, **167**, 318–51.
7. D'Appolonia, K. J. (1982). Health matrix – toxic waste isolation. *American Industrial Hygiene Association Journal*, **43**, 1–7.
8. Cope, C. B., Fuller, W. H. & Willetts, S. L. (1983). *The Scientific Management of Hazardous Wastes.* Cambridge University Press.
9. Anderson, A. (1984). Dry battery alarm in Japan. *Nature*, **309**, 576.
10. Benn, F. R. & McAuliffe, C. A. (eds; 1975). *Chemistry and Pollution.* London: Macmillan.

12

Protection against chemicals: will you wear it?

The protection of the individual against chemicals by physical means has two aspects. We commonly think of protection as being provided by equipment worn by the individual, such as special clothing or gloves. However, in an ideal situation this should be unnecessary. All protection should be afforded by the proper handling of chemicals in such a way that the individual is not exposed. This can be done by remote-handling techniques, and by ventilation or screening equipment that removes or excludes the chemical from that part of the work area occupied by the human operators. This ideal state is feasible in purpose-built, fixed installations, but may not be attainable in work at shifting locations. Therefore the two aspects of protection blend into one another. However, there should be a genuine attempt to adhere to the two principles of protection, which are listed below.

1) The workplace should be clean and free from chemical contamination, so that the individual worker does not have to wear protective equipment. This can be achieved by the proper planning of the handling and processing of the chemicals, and by the provision of adequate ventilation and dust-control equipment.
2) Individual protective equipment should be available for emergencies and rescue operations. It should not be relied on as the routine, day-to-day form of protection. If this seems to be necessary, look hard at the organisation of the work with a view to making the first principle effective.

There are cases where some modification of the two principles has to be accepted. For example, a low degree of personal protection is desirable in the workplace, to counter the unpredictable accident. This protection should include eye protection, work clothes (overalls, dust jacket, etc.) and sensible footwear. Also, there are many types of work in which the location is always shifting, so that the provision of protective installations is impossible. An example would be a contractor who is sanding and sealing wooden floors in private houses. He cannot have protection installed to shield him from either the dust of the sanding operation, or the solvent fumes from the sealer. Therefore he has to rely on personal protective devices (if he bothers with anything at all). The domestic handling of chemicals is another problem, which I will discuss as a whole later. There are thus three situations to be considered in relation to protection, as shown in Table 12.1. Firstly, we have the fixed location, which is the purpose-built installation in which protection has been (or ought to have been) considered as part of the original design. In real life, the function of the building may have changed, and air-filtration equipment or dust control measures may have been added later as add-ons, with consequent doubts about their efficiency. Secondly, there is the situation of the shifting location, exemplified by agricultural pest control activities, building work, etc. Here protection must largely be of the individual (coupled always with good handling practices). Lastly, there is the domestic situation, in which least protective gear is available, and in which least instruction can be given in safety matters. I will discuss the fixed installation, collective protection first, then the individual protection, and lastly the home situation.

It is profitable to think very carefully about the design of a chemical factory or laboratory with protection of the workers in mind, before deciding on the details of the processes to be performed. The most efficient or cheapest method of production may involve a chemical as an intermediate which is particularly toxic. In that case, is it better to substitute a safer chemical, lose slightly on efficiency but save on the costs of protection? The methods of operation need also to be considered.

Table 12.1. *Protective procedures appropriate to various work situations*

Situation	Protective procedures
Fixed location (factory, filling plant, refinery)	Protection included in original design and fitted as an integral part of the building; personal protection for emergencies only
Shifting location (agricultural spraying, home building, etc.)	Design equipment and plan procedures to reduce contamination of the operator to a minimum; use personal protection as necessary when other measures are not adequate
Home (detergents, cleaning preparations, pharmaceuticals, etc.)	Provide clear, simple instructions on handling and use; do not presume the existence of any protective equipment other than rubber gloves

Mixing and stirring in open vessels will generate high vapour and/or aerosol levels of chemicals, which may be avoided by use of another process. Similarly, loading of a powdered product through a hopper into bins will generate plenty of dust. Can this be done in another way? The point is that the need for controlling a hazard can be obviated by not generating the hazard in the first place. A little thought can save a lot of money. In practice, some problems will remain. With respect to chemicals, these are largely the problems of ventilating work areas to remove dust, fumes, and vapours. The technology for ventilation and filtration has been well examined and is available (1,2), so I will discuss it briefly here in relation to laboratory ventilation, with which I am familiar.

The most common arrangement for working with toxic chemicals is to have an enclosed, well-ventilated workspace or fume cupboard, in which the experimental apparatus can be manipulated from outside by an unprotected person. The fume cupboard is usually enclosed on three sides with impervious material (preferably stainless steel), with access from the front below an adjustable safety glass sash. The ventilation system for the fume cupboard has to be capable of maintaining an air velocity into the cupboard of at least 0.5 m/s, and preferably 1.5 m/s. The air flow through the exhaust fan is usually constant, therefore the air velocity will fall off as the sash is opened wider and the area for air entry is increased. An air velocity of 1.5 m/s with the sash 30% open (which allows for getting your hand inside) is a goal to aim at. The flow of air will be modified by your body outside the fume cupboard and by bulky bits of apparatus inside the cupboard, therefore air flows should be checked during a realistic rehearsal of an experiment. I have seen so-called fume cupboards with feeble air draughts which have spilled air outwards at the corners, because the incoming air has encountered a large central piece of apparatus which has caused a turbulent reverse flow. Remember also that the air has to come from somewhere. If the laboratory is small, or has a lot of fume cupboards, deliberate arrangements for supply air have to be made. Therefore you cannot expect air-conditioned comfort in the laboratory, unless you are prepared to spend a fortune conditioning the transient air.

What happens to the air exhausted from the fume cupboard? It is ducted to the outside and discharged through a stack at a sufficient height to clear neighbouring buildings. If the quantity of toxic chemicals being handled is small, the dilution in the air stream may be enough to render the discharge innocuous. If the quantity is large (and allowing for the possibility that it all may be released at once if you drop it) or very toxic, then the discharge must be filtered. Particulate filter systems for solid powders and non-volatile liquid aerosols, and absorbent filters (usually charcoal) for vapours can easily be included in the exhaust system. The required degree of treatment of the discharge has to be worked out in association with the regulatory authority responsible for the building. When treatment is deemed unnecessary, then this has to be justified in terms of likely maximum discharge concentrations compared to the regulatory limits. If you say nothing will be discharged, because you have fitted a filter system, then you find you are not believed until you explain in detail what happens to the effluent. On the receiving end, the nearby resident has every right to know what is being discharged near his home, therefore the design engineer and the installation manager need to be prepared to explain in clear terms what will happen.

Once all the vapours or fumes have been exhausted from the building you may feel that you can sit back and relax in a clean atmosphere. Not so. It is possible that the exhausted fumes may be drawn back into the building with the supply air. This depends on the geometry of the building, whether the exhaust stacks are high enough, the direction of the wind and many other factors. The best check is a practical test, as suggested by Bottomley (3). He showed that exhausted vapours did re-enter the laboratory building he was testing, but that the levels from any likely spill were not hazardous. Nevertheless he was able to make suggestions for improvements in the layout of stacks, air-inlet systems, etc.

It is unlikely that many of you readers are going to design fume cupboards. I have just used this as an example of the factors that need to be considered in designing good protective installations. Details of laboratory design are given in references 4 and 5. The design of an industrial chemical plant will involve the consideration of a large number of systems, infinitely more complex than this example. However, the basic message is still the same: money spent on good design at the start will save the management much expense later. Add-ons are expensive.

The methods of protection of the individual of course relate to the routes of entry to the body (Chapter 5). The important ones for us are via the respiratory tract and the skin, with the eyes as a minor but important route. Respiratory protection is inevitably considered separately from that of skin because the requirements are so different, but it is important to remember that the two types of protection have to be integrated. Respiratory protective devices (respirators, gas masks) can be classified as in Table 12.2, which outlines the main applications. The military respirator (Fig. 12.1) is the most advanced development of the full-face type. Air is drawn in through the filter canister when the subject inspires and flows first across the eyepieces, an action which helps to keep these free of mist. It is then inhaled through the nose. On expiration, the route through the canister is closed by a simple valve, and the stale air is forced out through an outlet valve, which is designed to seal during inspiration. The respirator has some form of simple diaphragm or voicemitter near the mouth to increase the clarity and volume of speech, which would otherwise be incomprehensibly smothered by the mask. The mask may also have some form of drinking tube to enable the wearer to get clean water from an enclosed water bottle.

Half-mask respirators cover the nose, mouth and chin only. The complexity depends on the filtering requirement. Disposable dust filters can be quite simple affairs (Fig. 12.2) whereas if the requirement is to remove vapours also, then the filters become more elaborate and bulky. One weakness of the half-mask type is the need to seal the mask against the skin across the nose. Quite often channels are left open down the sides of the nose through which unfiltered air can reach the nostrils.

The AIRSTREAM type mask is a new development which aims to reduce the stresses that other respirators impose on the wearer. Fresh air is actively supplied to the face under the visor (Fig. 12.3) by an electric motor and fan, and is drawn through a filter which can be particulate only, or particulate and vapour-absorbing. The air flow is sufficient to keep a slight positive pressure of air over the face, thus decreasing the chance of dust or

Table 12.2. *Classification of respiratory protective devices*

Category	Main uses
Filtering devices (dependent on purifying surrounding air)	
Full Face	When protection of the eyes and skin of the face is also required; used in conjunction with protective suits and hoods if skin contamination is a problem; military and industrial applications
Half mask (oronasal mask)	When protection of the respiratory tract only is necessary; minor industrial hazards
Airstream type	For maximum comfort and ease of breathing by wearer, coupled to moderate degree of protection; industrial use
Breathing apparatuses (Clean air is obtained from gas cylinder or other enclosed source)	
Open circuit (airline or self-contained types)	For use in high concentrations of toxic vapour, or when there is oxygen deficit in air; cleaning up chemical spills, and firefighting
Closed circuit	As above, but when reserves of clean air or oxygen are limited; various special applications

vapour intrusion through the gaps between visor and head. This type of mask is not as efficient a protection as a well-fitted, full-face respirator, but it has advantages for industrial use among moderate hazards. These advantages include absence of the resistance to breathing through a canister (since the air is supplied by fan), a cooling effect of the moving air over the face and a better field of vision. It is believed that these advantages will result in the user actually wearing the AIRSTREAM much more readily than a full-face respirator, so that the loss in theoretical protection is offset by the acceptance of the device by the wearer.

The wearer of the above respirators breathes the surrounding air, purified by passage through the filters. Clearly, if the surrounding air does not contain enough oxygen to provide for the requirements of the person,

Fig. 12.1 A modern military respirator. This is the British type S6 respirator. The filter canister is on the wearer's left; the 'snout' at the front contains the voicemitter and the outlet valve. Note the prescription lenses inside the eyepieces. The hood and chemical suit are also shown.

then this type of respirator is no good. Also, if the surrounding air contains a toxic vapour at so high a concentration, or of such a nature, that the filters cannot cope, then again this type cannot be used. We then have to use a supplied-air system, generically known as breathing apparatuses. These can be open circuit or closed circuit. The former type have fresh breathing air supplied from a reservoir, which may be remote from the wearer (who is connected by an airline) or it may be carried by him. Figure 12.4 shows a self-contained breathing apparatus, the underwater version of which has given the language a new word, 'Scuba'. The expired air is vented to the atmosphere. In closed-circuit systems the expired air is not vented, but is returned to an air-bag through a filter which absorbs the excess carbon dioxide. Oxygen is added to the bag to make up the deficit caused by respiration, then the air in the bag is available for rebreathing. The closed circuit systems are by their nature self-contained, there would be no point in a closed-circuit system on an airline.

Breathing apparatuses are used by fireman, since they may encounter

Fig. 12.2 A disposable half mask dust respirator. This model is the Gerson No. 1501, for use with non-toxic dusts such as sawdust, household and garden dusts.

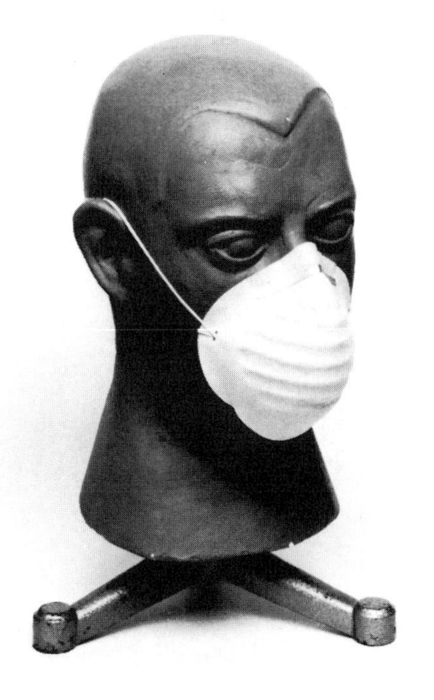

oxygen deficits, and by emergency rescue teams who have to enter areas of high vapour concentration, or of unknown risk. Filtering devices are used when the risk is lower and mobility is required. The selection of respiratory protection for a particular job needs careful evaluation. The equipment and its selection is adequately discussed in reference 6.

Respirators are perhaps the type of equipment from which the inexperienced person expects the most and understands the least. The first point that is not understood is that one type of respirator will not protect against all risks. A particulate filter will protect against asbestos fibre, but not against insecticide vapour. Further, an absorbing filter that will protect against insecticide vapour will not protect against hydrogen cyanide, because the charcoal in the filter is not a good absorbent for the small cyanide molecule. However, people go to a supplier and say they want a respirator. It is very difficult to extract from them any information on what they want protection against. The belief is almost universal that one respirator, any respirator, will protect against any risk.

Fig. 12.3 The AIRSTREAM respirator manufactured by Racal Safety Limited. At bottom right is the battery pack. The blower and filter are contained in the back of the helmet.

The next point is that given the appropriate respirator in perfect condition, protection is not assured. The equipment must be fitted to the wearer. There is a common belief that if the respirator is hung upside down around the face, it will give protection. Well, it will not. It has to be fitted. The preferred method is to fit the mask to the wearer and let an experienced person adjust the straps, explaining what he is doing and what is necessary in strap tension and positioning. Then the wearer is exposed in a chamber of tear gas. If leaks occur, the mask is readjusted. The wearer is made to move about so that he is assured that he has protection when moving on his job. Then he is asked to remove his mask before leaving the chamber. The sudden onset of tear gas symptoms demonstrates in the clearest way the protection the respirator has given him. Protection is thus a combination of good equipment plus good fit.

Fit leads to the contentious topic of facial hair. Long sideburns and/or beards lead to leakage around the peripheral seal of the respirator. The Australian Standard (7) requires that the face is clean-shaven, and limits

Fig. 12.4 A self-contained breathing apparatus. This model is the Normalair Compressed Air Breathing Set manufactured by Normalair Ltd.

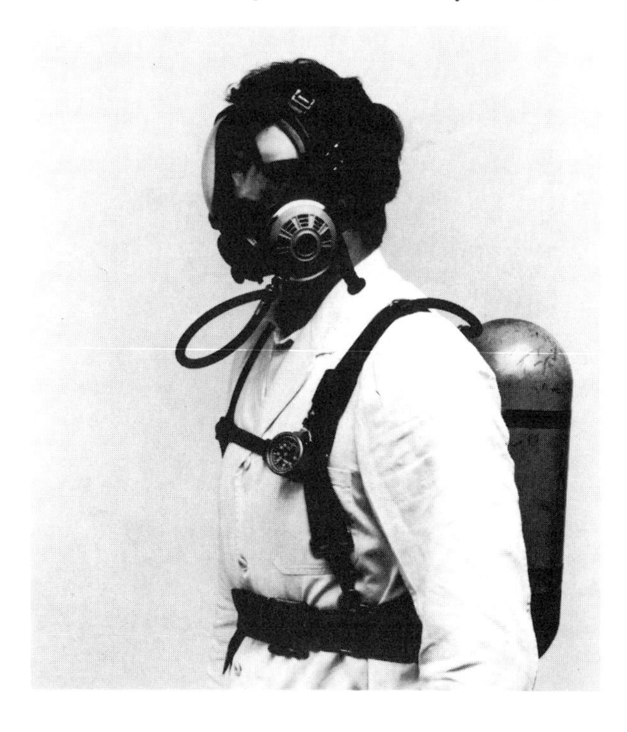

the extent of sideburns. Anyone not complying with this requirement cannot be fitted with a respirator; it is a waste of time and irresponsible to do so. We have had the occasional situation arise in which a bearded person has needed to wear a respirator because of his work, and we have refused to fit one. These situations have usually been resolved by proper counselling and discussion, the beard has come off. This is a management problem, not a technical one, and the successful outcome depends on the quality of management.

The next point to consider is that of maintenance. Equipment must regularly be examined and maintained. The pesticide operative's mask that has been hung on the back of his truck for 6 months in the sunlight and dust is unlikely to be of use to him. New equipment from the supplier also needs checking; never assume that you are buying a fully efficient system.

The respiratory tract is the part of your body most vulnerable to chemicals. A well-fitted and efficient respirator will give a very high degree of protection, but only if you take the matter seriously. Guides to the selection of respiratory protection appropriate to a particular need are available (e.g. in reference 7).

The skin is the other area of the body to be protected. Protection can be partial, as for gloves, aprons, overboots, or it may be total and integrated with respiratory protection, as shown in Fig. 12.5. Whether partial or total, the material selected for the garment must be chosen carefully to ensure that it really gives protection. Consider gloves, for example. Surgical rubber gloves are often suggested as giving protection against chemicals combined with the retention of manual dexterity. The protection is often illusory. They certainly protect against powders and aqueous solutions or suspensions, which is what the surgeon requires. They offer practically no barrier to many organic liquids, however, and in fact natural rubber soaks them up readily. In some ways rubber gloves may be worse than bare skin, for a liquid which has soaked into the rubber will be held against the skin for a prolonged period, thus allowing for greater skin uptake. To obtain long-term protection, thick gloves of butyl rubber are necessary. It may therefore be better to wear no gloves if delicate manipulations are necessary; frequent washing is then desirable.

Similar considerations need to be applied to materials for protecting the rest of the body. The material has to be impermeable to the particular chemical seen as the challenge. In practice, all materials are permeable in some degree. The combination of low permeability plus increasing thickness of material will increase the time that a chemical will take to penetrate through from outside to inside. Therefore protection can be expressed in terms of time; good materials are not impermeable, they act as barriers for longer than other fabrics. You must therefore decide for how long you want protection. Also, you must not don the same pair of gloves or overalls day after day; once contaminated their useful life is limited. These considerations apply to liquids only; in practice, any continuous barrier will protect against dry powder or vapours (unless the concentrations of the latter are massive). This topic is well and clearly covered in reference 8.

The person who is wearing protective clothing is encumbered in several ways. Gloves which really protect have to be thick and hence manipulations are clumsy; clothing and equipment slow movement; vision may be

Fig. 12.5 A supplied air breathing apparatus with integral protective suit. This wearer is obviously very happy despite the cumbersome appearance of the equipment.

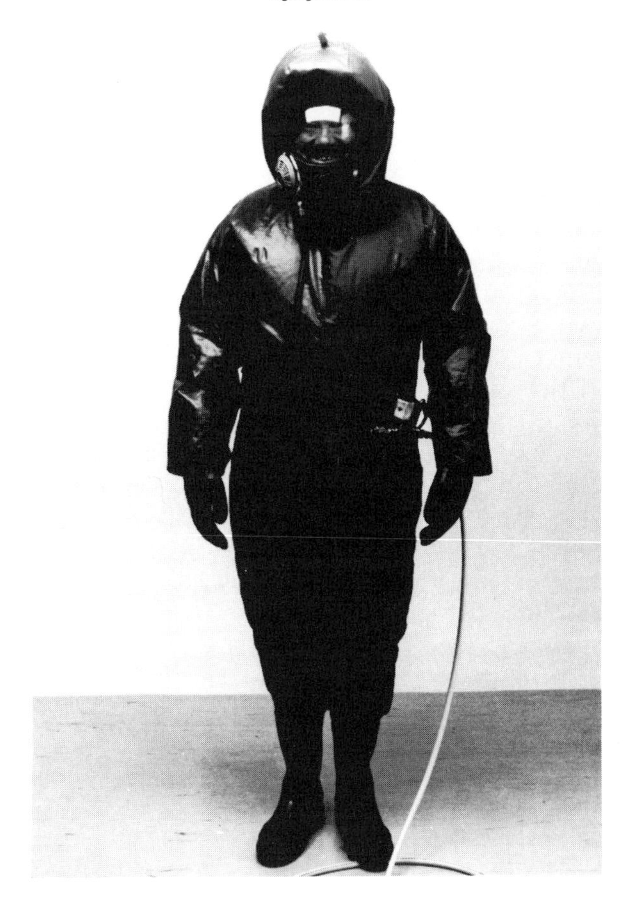

restricted by the respirator. Perhaps the major problem for the wearer of full protection is the fact that total encapsulation in a protective envelope prevents the loss of body heat. This is not a problem only for persons in hot climates, for work in temperate or even cold climates will generate body heat that has to be lost. The rate of loss depends on the insulating properties of the clothing and the temperature gradient between body and the environment. When the temperature of the latter approaches that of the body (37 °C), then the main mechanism for loss of body heat is by sweating. However, as you have anticipated, protective clothing will partly or wholly prevent the evaporation of sweat. There is therefore some sort of relationship between the three factors of rate of working (which determines the heat produced), the surrounding climate and the period for which a particular type of protective clothing can be worn before the wearer collapses.

It is the total encapsulation of the wearer that is the major factor, and variations in the clothing or in the climate are less important determinants, so that as a reasonable simplification it is the rate of working that determines the time that protective clothing can be worn. The whole topic is well discussed in reference 9. It is important because heat stress is a serious condition for a person, which can very suddenly be fatal. When you work in hot conditions, you feel unpleasantly hot and sweaty from the start. You do not know if your body temperature is 38 °C (no cause for alarm) or 39 °C (becoming dangerous) or 39.5 °C (very dangerous). At the latter temperature you have probably collapsed and unless you are cooled very quickly your body temperature will continue to go up and above 40 °C death is likely. I will not go into the details of heat stress here, but wish to emphasise the seriousness of it in relation to protective clothing. If you have to do a job in full clothing, or supervise other people doing the job, you must consider the possibility of heat stress and make an estimate of the safe working time, taking into account all the variable factors. There is quite a literature on the topic, and reference 9 will be a start.

The body can be cooled effectively by moving air circulating between the body and the protective clothing. Supplied-air systems make use of this by allowing a proportion of the air to enter this space, and produce cooling by convection and the evaporation of sweat. This is a further advantage of such systems, and therefore despite their apparent complexity and the limitations of having a hose to trail around, they are often the only protective equipments that can be used in conditions of high chemical risk and hot working temperatures.

The physical limitations of protective clothing (which I have summarised in Table 12.3) lead to the topic of the psychological factors which are involved in getting people to accept such apparel. The problem is, since any kind of protection affords some physical limitation or inconvenience, then the temptation is not to wear the piece of gear in question. The gloves are sweaty and clumsy; throw them away. The protective goggles mist up and also pinch your nose; don't wear them. The respirator makes your face feel hot and seems to impede your breathing; take it off, or slide a pencil over your ear and under the edge of the respirator seal so that air enters more easily from the outside. These situations arise all the time. In fact, the technology of protective clothing has two equal parts, that of the protection afforded by the equipment, and that of the ability and willingness of the person to wear it. One is the science of materials chemistry, the other is that of human physiology and behaviour. Never fall into the trap of believing that protection is all about the former, and nothing about the latter. A little experience in protection will soon show you that most of your problems are in the latter area.

We have already considered the problem of facial hair as a factor incompatible with the wearing of a respirator. Other aspects of respirators are discussed by Morgan (10), who points out that whilst the technical design of equipment has progressively improved, there has been little effort to investigate the limitations of the wearer. His paper is therefore more a plan for research rather than a review of knowledge. This attitude is justified. There is an awful lot of anecdotal information possessed by those who have worked with respirators for years, but very little that is documented in a systematic manner. A fertile source of information would

Table 12.3. *Problems associated with the wearing of protective clothing*

1. Lack of dexterity	Fine manipulations impossible; bodily movements awkward
2. Heat stress	Reduces rate and duration of working; can produce a fatal collapse
3. Limitation of vision	Peripheral vision is limited; objects cannot be brought close to eye (e.g. objectives of telescopes or microscopes)
4. Increases breathing resistance	Inspiring through a filter or breathing against a valved system gives an impression of increasing respiratory work; often more apparent than real
5. Limitation of ability to communicate	Breathing equipment muffles voice, also impedes other visual signs and facial expression
6. Feeling of isolation and lack of personal identity	A psychological problem arising from factors 3 and 5, and the difficulties of recognising others and being recognised oneself when all are in protective clothing

be from military training, as many armies train actively with gas masks and thus have pools of subjects ready. The situation is not different for clothing as opposed to respirators; there is very little research done on the psychology of protective apparel.

If we leave the theoretical aspects of this subject and consider the person at work on the job, the problem becomes a management one. The boss has to know how to make sure that protection is made use of. This can involve a mixture of direct order and enforcement with instruction and explanation. It is probably best to rely largely on the latter, even if the workforce is perceived as being dull-witted. Certainly the unions or other worker representation should be involved in discussions on the need for protection. They may well ask for measures that are unnecessary; they may also be asking for protection that is necessary, but that the management privately feels is too expensive. These matters have to be argued to a compromise. However, it is in the interests of both parties that the worker uses what protection is available (as in Fig. 12.6); surprisingly, often the worker does not share that view.

Fig. 12.6 Work in a chemical factory can be clean, as shown here. The work area is free from dust and litter, and the workers wear a sensible degree of protective clothing (overalls and helmets) which does not encumber or aggravate them (photograph courtesy of ICI Plc).

The degree of responsibility of management can be gauged by its attitude towards the retention of a permanent workforce. If it is happy to hire and fire an ever-changing population of employees, then it most likely doesn't give a damn about protecting them. Conversely, if the employer builds up a stable group of workers, then it is in his or her interest to protect them, for any compensation claims will come right back to the known employer. The use of individuals as contractors or independent operators by a central firm (common in pest control) is an insidious practice which allows the management to distance itself from its responsibilities of protecting its workforce.

If we now consider the domestic situation, it seems to me that it is unreal for chemical manufacturers to expect the householder to have available any protection other than household rubber gloves, and overalls and aprons. Therefore chemicals for sale on the domestic market should be capable of being handled safely by someone wearing this basic equipment. This would then particularly limit the domestic use of toxic substances which are either volatile or used as aerosols. However, from another point of view, I recommend that further equipment is available at home, principally eye goggles and disposable half-masks to protect against dust particles and coarse aerosols. Certainly sensible protection, which can be quite cheap and readily obtained, would help to reduce the incidence of poisonings in the house, garden and farm. The most blatant example of lack of care that I can recall was that of the man who sprayed the inside of his glasshouse with insecticide in hot weather when dressed only in shorts. He went to hospital, but I cannot remember if he died.

To summarise the above, there is a technical aspect to protection which has to be thoroughly understood, and its limitations realised. In addition to this, there is the complex question of the human who has to wear the equipment, and his attitudes, positive or negative, towards it. Even if you, the reader, never have to wear protective equipment, you need some understanding of the philosophy of its use to be able to comment on industrial accidents involving chemicals, and to require sensible legislation on protection from your parliamentarians and officials.

References

1. Russell, D. K., Keeny, Q., Mutchler, J. E. & Clayton, G. D. (1973). Design of ventilation systems. In *The Industrial Environment – its Evaluation and Control*, pp. 609–28. Washington: National Institute for Occupational Safety and Health.
2. White, P. A. F. & Smith, S. E. (Eds.) (1964). *High-Efficiency Air Filtration*. London: Butterworths.
3. Bottomley, G. A. (1983). Does fumecupboard waste air significantly contaminate other rooms? *Chemistry in Australia*, **50**, 198–201.

4. Ferguson, W. R. (1973). *Practical Laboratory Planning*. London: Applied Science Publishers.
5. Everett, K. & Hughes, D. (1979). *A Guide to Laboratory Design*. London: Butterworths.
6. Ballantyne, B. & Schwabe, P. H. (Eds.) (1981). *Respiratory Protection. Principles and Applications*. London: Chapman & Hall.
7. Anonymous (1982). *Selection, Use and Maintenance of Respiratory Protective Devices*. Australian Standard 1715–1982. Sydney: Standards Association of Australia.
8. Schwope, A. D., Costas, P. P., Jackson, J. O. & Weitzman, D. J. (1983). *Guidelines for the Selection of Chemical Protective Clothing*, Vol. 1. *Field Guide*. Vol. 2. *Technical and Reference Manual*. Cincinnati: American Conference of Governmental Industrial Hygienists.
9. Leithead, C. S. & Lind, A. R. (1964). *Heat Stress and Heat Disorders*. London: Cassell.
10. Morgan, W. P. (1983). Psychological problems associated with the wearing of industrial respirators: a review. *American Industrial Hygiene Association Journal*, **44**, 671–6.

13

The factual basis for concern about herbicides

It is very easy in a book of this kind to produce a superficial survey of a broad topic, without ever getting down to details. I have selected one topic to consider in depth, that is from all aspects including technical, social, political, informational, etc. Without doubt the subject of the greatest public discussion about chemicals has been the herbicide 2,4,5-T, and the related topics of 2,4-D, dioxin and Agent Orange. Therefore I wish to devote a chapter to the factual basis of this issue, and then follow with a second chapter which discusses the public's perception of the issue. However, I will introduce the topic with a consideration of another chemical, asbestos, which is not synthetic, but mined and used without chemical modification.

My reason for first discussing asbestos is that I believe one feature of its toxicity has been the stimulus for public alarm about many other chemicals, including 2,4,5-T. This feature is the long latent period between exposure to asbestos and the appearance of disease, which period may be 20 to 30 years. If a workman falls into a vat of sulphuric acid, he is physically consumed in a few seconds. The accident is ghastly, yet cause and effect are clear; the hazard is well defined. If a worker collapses after inhaling solvent fumes in a confined space, she can be rescued and will in most cases recover quickly. She may suffer long-term debilitating effects of the accident so, in this case, the outcome of the chemical exposure is less defined than in the former case, but the cause is again obvious. However, if a member of the public is exposed to a chemical for a period during his or her life (whether at home or in the factory or anywhere else) and many years later develops a disease, what then? The development and outcome of the disease cannot be predicted as the medical profession has no prior knowledge of it, except in general terms relative to conditions with similar symptoms. Further, the cause is difficult to establish, or may be impossible to assign to a particular chemical if that chemical was used for a short time in a process, then replaced with a more effective one.

The only method of establishing a link between cause and effect is by retrospective epidemiological studies. For this to be effective, the factors shown in Table 13.1 are required. If the disease is distinctive, has occurred with increasing frequency and only occurs among workers in one particular industry, then it is easy to connect effect with cause. If any of these desirable features are absent, then we may not know about it at all. There will be a very slight increase in the incidence of disease in general, too small to show above the normal statistical fluctuations due to many other causes (Chapter 7). Even if these features are present, it may be a long time before the cause and effect are well recognised. For example, consider bladder cancer and asbestosis as discussed in Chapter 8.

The background to the asbestos story is summarised in reference 1, and I have already mentioned some aspects related to cancer (Chapter 8). Here I will just discuss those features which relate to the recognition of the problem, on the lines of Table 13.1. Mesothelioma is a rare disease but of an unusual form and thus easy to discern and record. It could then be linked to the one specialised industry (asbestos processing) that was its cause. The lung cancers induced by asbestos would be much less obvious in a collection of general statistics. The incidence of asbestosis and lung cancer alone would not have been sufficient in statistics of the general

Table 13.1. *Factors which assist in recognition of a chemically induced disease*

Factors	Remarks
Recognition of disease as novel	
Distinctive features of disease	These factors make the disease discernible in medical statistics
Relatively high incidence	Must be someone alert enough to recognise novel nature of the disease
Link to cause	
Personal histories of patients	May be difficult if patients have died
Occurrence in one particular industry or environmental situation	Good link to cause, if it exists
Nature of suspect chemicals clearly defined in technical terms	Patient may only know a brand name, or no name at all; chemical preparation may have been a mixture of varying composition
Known exposure to the chemical, including dose (concentration and time of exposure)	Best evidence to establishing a link, but less likely to be attainable as exposure recedes in the past

population to be remarkable. The personal histories of patients were useful in establishing a link, for they were working with asbestos as the main feature of the industry, not as an incidental chemical in some process, and had in many cases been in the industry for years. The actual degree of exposure in most cases was unknown, but it was certainly heavy. The nature of the chemical to which asbestos workers were exposed is interesting. The link between mesothelioma and asbestos was originally with blue asbestos (crocidolite). However, it is now clear that lung cancer is caused by all three types of asbestos (crocidolite, chrysotile and amosite), and this may be so for mesothelioma too, although crocidolite is the most potent cause of this disease (1). Do not forget that smoking increases susceptibility to asbestos-induced disease. The features of the asbestos story that aided the definition of effect and cause were the distinctive nature of mesothelioma and the circumscribed but heavy industrial exposure. The different properties of forms of asbestos had a slight confusing effect.

From a technical point of view the history of asbestos illustrates nicely the difficulties of linking cause with effect, even when conditions are favourable. The journals are still full of epidemiological studies on people who worked in the asbestos industry 10 or more years ago so that the picture is not yet totally defined. But for the general public, I again emphasise that the major product of asbestos has been the fear of diseases which strike years after exposure. Asbestos-induced disease takes 10 years or more to occur and is not detectable epidemiologically until after 15 years or more. The one death of which I have had personal knowledge caused by a defined exposure to asbestos, occurred 26 years after that event. In the mind of the public, this is the most worrying and unsettling fact – have I already been exposed to a chemical that will kill me in 10 years time? The scientific community has no answer to this emotive question.

Asbestos sets the scene for 2,4,5-T. Let us be clear, firstly, about the relationship between the chemicals in the 2,4,5-T group. The herbicides 2,4-D and 2,4,5-T (see Fig. 2.3 for chemical structure) belong to the class of chlorophenoxyacetic acids. Commercial preparations of 2,4,5-T are contaminated (in varying degrees) with dioxin, a highly toxic chemical. Dioxin does not occur in 2,4-D. Agent Orange was the code name given to a mixture of the butyl esters of 2,4-D and 2,4,5-T. The formulation as butyl esters rather than free acids meant that the preparation was oil soluble rather than water soluble. These, then, are the actors in the drama.

The question of whether or not there is a significant toxicity associated with 2,4,5-T and related chemicals is a very interesting and complex one. Its ramifications extend from chemistry to industrial hygiene and to social questions such as the role of the media in determining society's attitudes to unresolved scientific debate. The subject can be divided into three topics.

1) The intrinsic toxicity of the herbicides 2,4-D and 2,4,5-T. Since 2,4,5-T is always contaminated to some degree with the impurity dioxin, known to be toxic, the intrinsic toxicity of 2,4-D (not subject to dioxin contamination) is less complex to study.
2) The toxicity of dioxin present as an impurity in 2,4,5-T, TCP (Trichlorophenol) and similar chemicals, considered in relationship to the very low concentrations in commercial preparations, and the very much higher concentrations in wastes from chemical plants.
3) The effects attributed to the use of Agent Orange as a defoliant in Vietnam.

The herbicide 2,4-D was developed during the Second World War as an attempt to mimic the natural plant hormones and, in fact, its success as a destroyer of plants results from its over-effectiveness as a growth hormone. The chemical is particularly useful on cereal crops, as the narrow-leafed plants are not harmed but broad-leafed weeds are killed. In a post-war world very short of food the chemical found instant acceptance; production and use increased rapidly. There is therefore a history of 35 years of intensive use of 2,4-D and its related chemicals; 2,4,5-T came into general use in 1948. In the latter years of this period there have been increasing numbers of reports which associated the spraying of 2,4-D and/or 2,4,5-T in an area with the subsequent development of a health problem in people living close by. The most common health problem mentioned is the production of deformities in children before birth. In all cases in which the alleged effects of 2,4-D have been investigated fully, there has been little evidence to implicate this herbicide as a causative factor. We must note that 2,4-D has low-to-moderate acute toxicity of 100–800 mg/kg (as measured on various mammalian species); this is about the same as for aspirin (see Tables 4.1 and 4.2). The argument is not about this, but about the effects of very low concentrations of 2,4-D in the environment.

If we now turn from 2,4-D to 2,4,5-T, the issue becomes greatly complicated by the inevitable presence of dioxin in varying amounts in the 2,4,5-T preparations that have been used. Nevertheless, despite repeated claims that the use of 2,4,5-T has been followed by an increased incidence of birth defects or other health problems in the area that had been sprayed, none of these claims has been substantiated. In the case of epidemiological studies of possible 2,4,5-T effects, only in one set of studies in Sweden (2) has evidence been accumulated that the use of 2,4,5-T has been harmful to spray operators.

The toxicities of 2,4,5-T and dioxin can be determined in animals with reasonable accuracy. The acute toxicity of 2,4,5-T has been reported to range between 300 and 1000 mg/kg (LD_{50}) when given by mouth to various animal species (3). Note firstly that we do not know the dioxin content of the 2,4,5-T preparations used in these tests (in 1954 it could have been quite high), secondly we do not know if man is more or less susceptible to 2,4,5-T than the rabbits, chicks and rodents used in the tests. Thus 2,4,5-T toxicity falls between the moderately toxic and the slightly toxic classes of Table 4.2. Unfortunately, I have been unable to find any recent figures for 2,4,5-T acute toxicity, related to known, low levels of dioxin contamination; one would expect the toxicity to be less. Equivalent acute toxicities for dioxin are 0.6 (guinea pig) to 300 μg/kg, with some figures to 1 mg/kg for the hamster (data summarised by Hay in reference 4). In other words, dioxin as a single dose is at least 1000 times more toxic than 2,4,5-T, probably more than 10 000 times. Note the great variation in effect on different species of animals. Hence the importance of low dioxin concentrations in commercial preparations of 2,4,5-T. When considering the 2,4,5-T figures, you should remember that you would need to take at least 30 g (one ounce) by mouth of the active ingredient, not a diluted spray formulation, to have a 50% chance of killing yourself.

The public concern is, of course, more about chronic effects of 2,4,5-T and dioxin, than with acute effects. A number of studies (summarised in reference 5) done in recent years when the contamination by dioxin would be low, give a general picture of chronic effects of 2,4,5-T on animals. A dose of about 30 mg/kg/day can be sustained for long periods (up to 2 years) with few deaths but with marked pathological changes. There is great variation around this level, with species of animal and with the conditions of the experiment. The pathological changes seen were those expected from a general metabolic poison, e.g. damage to liver and kidneys principally. No mention is made of tumours. A summary of evidence as of 1982 (6) found inadequate evidence for the human or animal carcinogenicity of 2,4,5-T (and 2,4-D).

Chronic exposure to dioxin, at levels around 1 μg/kg/day, does produce marked effects on animals and the effects (summarised in references 4 and 5) are related to the formation of cancers, interference with the immune system, and chloracne (of which more later). The IARC (6) study finds sufficient evidence for dioxin carcinogenicity in animals, but inadequate evidence for the property in humans (probably only a reflection of the difficulty of getting good evidence for humans). Dioxin is therefore quite a different proposition from 2,4-D or 2,4,5-T.

Well, where are we now? The herbicides 2,4-D and 2,4,5-T have intrinsic toxicities of moderate to slight, by single dose. Chronic effects are not great. By contrast, dioxin is very toxic, acutely or chronically. Clearly interest focusses on the dioxin content of 2,4,5-T, now largely historical, as the aim now is to keep the dioxin content of 2,4,5-T below 0.01 ppm. This is 1 μg in 100 g of 2,4,5-T; if we estimate the human toxicity (acute) of dioxin as 10 μg/kg, then the traditional 60 kg adult human is going to need 60 kg of the preparation for death to occur (by the dioxin effect). This little exercise is just to show that 0.01 ppm is a very low contamination level.

We can now return to the human situation again, having for the present got what information we can from the animal experiments. As I said before, the result of numerous public inquiries into the claimed harmful results of 2,4,5-T spraying programme has been to find no involvement of 2,4,5-T in whatever health problem had occurred. Because these conclusions are quite contrary to public belief, I will list some of the reports. In the United Kingdom, the matter has been reviewed several times. No sound medical or scientific evidence was found that harm would result from the continued use of cleared 2,4,5-T herbicides for recommended purposes in the recommended way (7). An investigation in Australia into a group of birth abnormalities supposedly linked to the spraying of 2,4-D or 2,4,5-T in the Yarram district found that there was no evidence for any such link (8). Similar conclusions were reached about three clusters of birth defects in New Zealand (9). The history of 2,4,5-T has been more chequered in the USA, partly because usage has been high and because the Department of Agriculture and the Environmental Protection Agency have responded quickly to publicly expressed concerns.

Severe restrictions were placed on the domestic and agricultural usage of 2,4,5-T in the USA in 1970, at about the same time as defoliation in Vietnam was stopped. This action arose from the growing public pressure that was being exerted on governmental agencies as a result of increasing awareness of the significance of contamination by dioxin, and of the reports that 2,4,5-T and 2,4-D produced birth defects in experimental animals. Then followed a long period of discussion and regulatory hearings on the use of 2,4,5-T. A further crisis occurred in 1979 with the submission of the Alsea II report (10). It is worth looking more closely at the Alsea situation.

The Alsea area of Oregon, USA, is a forested area bordering on the Pacific Ocean. It is part of the ancient and worn Coastal Range which is well timbered and watered by many streams and rivers resulting from the heavy rainfall. Most of the human population lives in small towns on the coast, with the remainder living along the river valleys. I remember it as a very pleasant area, memorable for Pacific salmon and steelhead (seatrout) fishing. In 1978 the women of this area noted an apparent connection between the seasonal use of 2,4,5-T in the forests and the number of spontaneous abortions that occurred. This was investigated by the Environmental Protection Agency (EPA) who initially found no clear connection between the two phenomena, but later decided that there was (10). Their main observation was that the number of abortions recorded rose to a peak in the month of June, when data for the years 1971–7 were averaged month by month. This was correlated with seasonal 2,4,5-T use. A third report from another source (11) criticised the methods of the Alsea II report (10) and concluded that no correlation existed between spontaneous abortion in the Alsea area and 2,4,5-T use. Some of the criticisms centred on the treatment of data; for instance the abortion peak in June only occurred in the averages because of a very high value in June 1976. This very high value was in fact 10 – leading to the further criticism that such low total numbers were involved as to render meaningful analysis difficult. Other criticisms were based on the failure of the EPA to describe accurately the populations examined. Thus the Alsea population is predominantly urban, not rural. Spontaneous abortions occur for a wide variety of reasons in any population and show erratic variation due to chance if small populations are examined. The Alsea reports in total prove nothing except the difficulties of this kind of study. They tell us nothing about 2,4,5-T.

At this point you should refer back to the discussion on clustering in Chapter 7. Since 2,4,5-T has been used very widely throughout the world, it is evident that its use will be associated very occasionally with a cluster of victims of a disease. That the disease reported is malformation of the foetus is a result of the popular preconception that 2,4,5-T causes such an effect; these are the effects people are watching out for to report in association with herbicide usage.

The fate of 2,4,5-T thus hangs uneasily in the balance in the USA and elsewhere. Current regulations stipulate a maximum dioxin level of 0.01 ppm, which largely removes the original grounds for banning 2,4,5-T. However, the only grounds for concern about the use of these two herbicides are the Swedish studies of Hardell and colleagues (2), which I referred to before. These studies cannot be neglected. However, they need confirmation before they can be accepted as conclusive evidence, given any study of that kind with limited numbers of persons involved is subject to random statistical fluctuations which can push a conclusion in either direction – confirmatory or inconclusive.

We now have to consider directly what has become the main cause of public worries about 2,4,5-T and what has initiated most of the public inquiries, namely the fear that these chemicals can cause birth defects in children when the mother is exposed. Is 2,4,5-T a teratogenic agent? The answer is probably yes, because almost any chemical is. We have to remember the all important dose–effect relationship (Chapter 4). The human embryo is most susceptible to chemical influences between days 18 to 60 after conception, when the main organs and structures are being developed. Large amounts of any chemical may interfere in this process, resulting either in abortion of a grossly deformed foetus, or in the birth of a child with a lesser deformity. A large number of studies (5) has demonstrated that 2,4,5-T (with dioxin as contaminant) and dioxin are teratogenic in rats and mice, the deformities mainly being cleft palate and kidney impairment. Sheep and monkeys seem less susceptible, although the data are insufficient to allow firm conclusions. It is not clear how teratogenic pure 2,4,5-T is. The dose levels of commercial 2,4,5-T that produce birth defects in animals are around 50 mg/kg daily, fed on days 6 to 15 postconception. The equivalent dose of dioxin is about 2 μg, i.e. it is 25 000 times more effective. Once again, therefore, any problems in the human situation in the past have most likely been with dioxin contamination at relatively high levels in 2,4,5-T.

I have covered the first point I set out to consider (page 170), and have largely covered the second point on dioxin. As regards the other aspect of dioxin, that is the effects of relatively large concentrations of dioxin in waste from chemical plants and in the effluent from factory explosions (e.g. Seveso) I refer the reader to the book of Hay (4), which discusses all aspects of dioxin very adequately, and to Chapter 11 of this book. I will just point out that dioxin is without doubt a very toxic chemical, but that the actual quantities that have been produced have always been low. For this reason, it is probably only the waste dump and factory accident situations that need cause concern to the public. One other feature of dioxin is that it carries its own telltale trademark when it has selected a human victim. This trademark is the development of chloracne, a persistent and disfiguring form of blackheads (12).

We now turn to the third of the points I wished to consider, the even more controversial issue of Agent Orange. When considering Agent Orange, I must make it clear that at this time no one can write a conclusive assessment of its alleged effects. We do not have all the data, and we are far too close in time to make a dispassionate judgement. 20 years from now the final appraisal may be made. At present I will struggle with this subject to produce as good a summary as current information allows. The basic facts are that during the Vietnam War the United States sprayed large areas of forest with Agent Orange and other herbicides to destroy the cover that the trees provided to Viet Cong soldiers. Agent Orange contained 2,4-D and

2,4,5-T. In addition to the general spraying, limited areas received much higher concentrations when the planes were occasionally forced to dump the load due to mechanical faults or military action. This activity resulted in the exposure of Vietnamese (military and civilian), US, Australian and other allied military personnel to varying concentrations of the Agent Orange. To this exposure has been attributed a host of medical problems in Vietnamese civilians and in US and Australian veterans of the war. The major concern is that Agent Orange has caused congenital defects in children subsequently born to men who served in Vietnam. This is not the only reported ill effect; many psychological problems are ascribed to Agent Orange exposure.

The Agent Orange (and related herbicides) used in Vietnam contained varying concentrations of dioxin, up to 50 ppm or more, although later samples used in larger volume in Vietnam may have had less. Thus, of Agent Orange (and Purple) stocks left after spraying stopped in Vietnam (measured in 1972), samples from Johnston Island contained a mean concentration of 1.91 ppm, and from Gulfport, 1.77 ppm (13). It is estimated that 167 kg of dioxin were distributed over Vietnam in 20 000 tonnes of 2,4,5-T, which was part of the total 48 500 tonnes of herbicides sprayed from the air.

It is impossible to estimate the dosage received by the veterans individually. Soldiers who passed through sprayed areas after it had settled would receive very little. Others claim to have been soaked with the spray as the aircraft went overhead. The group with the most intense and longest-term exposure would be the men handling the spray in bulk, refilling the aircraft tanks and cleaning up afterwards. For this reason, the veterans from this operation (Ranch Hand) were selected by the US Government as subjects for an intense investigation to find if any health problems had arisen from their exposure to Agent Orange. A preliminary report in February 1984 (14) concluded that there was insufficient evidence to support a cause and effect relationship between herbicide exposure and adverse health. This conclusion is tentative; the study found many observations of a medical nature that required further investigation. Chloracne which, as we have seen, is a sensitive indicator of exposure to dioxin, was not found in this study, nor does it feature commonly in the claims presented by the veterans. This inquiry did report a slight increase in minor birth defects (birth marks, etc.) in children of the Ranch Hand operators as opposed to children of a control group of veterans. This finding was made on the basis of self-reported data, which require confirmation by more objective means (medical records, etc.).

The Australian Government is similarly engaged in a series of studies, which so far have not linked service in Vietnam with any health problems that might be due to herbicide exposure. In fact one study (15) reports that there is persuasive evidence that Vietnam service has not been associated with any important increase in the risk of birth defects in children of veterans. Further work in Australia was associated with a Royal Commission on the effects of exposure to chemicals on the health of servicemen in Vietnam, which reported in August 1985 (16).

The commonest claim made by Vietnam veterans is that they have sired deformed children after having been exposed to Agent Orange in Vietnam. The implication is therefore that the herbicide (or dioxin) has effected some change in the germ cells of the father, which has resulted some years later in defective sperm fertilising the egg and producing a defective embryo. Such results have been induced experimentally in animals by chemicals which are strongly mutagenic, to which class 2,4,5-T and dioxin do not belong. Alternatively, it is conceivable that a teratogen like dioxin can be stored in the male body for the latent period and somehow transferred in the seminal fluid to cause a defect in the embryo. However, a review by Pearn (17) concludes that there is no positive experimental evidence to support this theory, nor any evidence from clinical studies. Male rats given near-lethal doses of dioxin were allowed to mate. No effect of dosage was found on the litters they sired, despite the obvious toxic signs on the fathers and the short time interval between exposure and conception (18). No one has yet found excessive malformations in the children of veterans, either.

In objective terms, there is no evidence yet that exposure to Agent Orange has caused general health problems in veterans. Many investigations are still to be completed in the USA and Australia. The question is, how much public money should be spent on further enquiries, if up to now no evidence has been found to support the claims of veterans?

I hope the reader has followed through this chapter, because although the factual basis is dull, no one can claim to hold a meaningful opinion unless they have digested the facts. In summary, what have we found? The chlorophenoxyacetic acid herbicides and the contaminating dioxin have been subjected to an enormous amount of toxicological research. The data are summarised in Table 13.2. It must be realised that this is a crude summary of a very complex situation; go to the original sources if you wish to study this in detail. The toxicity of 2,4,5-T is that of a slightly toxic to moderately toxic chemical (Table 4.2) which means that it is comparable to thousands of other agricultural and industrial chemicals. By contrast, dioxin is extremely toxic; in the class of biotoxins or supertoxic chemicals, depending on which animal species we test it on. As regards chronic toxicity or the production of birth defects, dioxin is again much more potent than 2,4,5-T. Therefore the toxicity of commercial samples of 2,4,5-T is largely determined by dioxin content; the current standard of 0.01 ppm (1 in 10^8) dioxin in 2,4,5-T reduces the problem of contamination below any practical importance. The intensive use of 2,4,5-T and 2,4-D over 35 years has not resulted in any general health problem among spray operators or the public. This statement may need qualification after further examination of the Swedish reports and the results of the various Agent Orange inquiries. What we do know quite positively is that 2,4,5-T usage has not caused a health problem that could be compared, say, to the risks of driving a car, smoking cigarettes, or becoming a chronic consumer of aspirin or alcohol. If any effect is confirmed, it will be the palest shade of grey (Chapter 4).

We survey the chemical world with a feeling of bewilderment. What the media and the public see as a great chemical evil is reported by scientists as being a useful chemical which has been used on a large scale for many years without causing any health problems at all. We are right to be bewildered. Are the media the tools of pressure groups that wish to distort facts for some devious reason, and is the public a hysterical body of alarmists not interested in facts? Or are the scientists and government officials in many countries consistently producing fallacious information in support of the

Table 13.2. *Summary of data on the toxicity of 2,4,5-T and related compounds*[abc]

Chemical	Acute toxicity (LD_{50}, mg/kg)	Chronic toxicity	Teratogenic doses (days 5–15 of pregnancy in mice)
2,4,5-T (with dioxin in varying concentrations)	300–1000	30 mg/kg/day for 2 years causes few deaths but obvious signs of poisoning	50 mg/kg
2,4-D	100–800		
Dioxin	0.0006–1 (0.6 μg–1 mg)	1 μg/kg/day produces marked effects, including cancers	2 μg/kg

[a] The figures are the results of experiments on various animal species, and are averaged from many results

[b] Agent Orange was a 1 : 1 mixture of 2,4,5-T and 2,4-D, in the butyl ester form. The toxicity of this mixture can be estimated from the above figures.

[c] The origins of the data are given in the text.

chemical industries of the world? We will try and answer some of these queries in the next chapter.

References

1. Wagner, J. C., Berry, G. & Pooley, R. D. (1980). Carcinogenesis and mineral fibres. *British Medical Bulletin*, **36**, 53–6.

2a. Hardell, L. & Sandström, A. (1979). Case control study of tissue sarcomas and exposure to phenoxyacetic acids and chlorophenols. *British Journal of Cancer*, **39**, 711–17.

2b. Hardell, L., Eriksson, M., Lenner, P. & Lundgren, E. (1981). Malignant lymphoma and exposure to chemicals, especially organic solvents, chlorophenols and phenoxy acids: a case control study. *British Journal of Cancer*, **43**, 169–71.

3. Rowe, V. & Hymas, T. A. (1954). Summary of toxicological information of 2,4-D and 2,4,5-T type herbicides and an evaluation of the hazards to livestock associated with their use. *American Journal of Veterinary Research*, **15**, 622–9.
4. Hay, A. (1982). *The Chemical Scythe. Lessons of 2,4,5-T and Dioxin*. New York: Plenum Press.
5. Hayes, M. K. (1981). *The Health Effects of Herbicide, 2,4,5-T*. Washington: American Council on Science and Health.
6. IARC (1982). *IARC Monographs on the Evaluation of the Carcinogenic Risk of Chemicals to Humans. Chemicals, Industrial Processes and Industries associated with Cancer in Humans*. IARC Monographs, Volumes 1 to 29. Supplement 4. Lyon: IARC.
7. Advisory Committee on Pesticides (UK), (1980). *Further review of the safety for use in the UK of the herbicide 2,4,5-T*. London: Ministry of Agriculture, Fisheries and Food.
8. Department of Primary Industry (1978). *Report of the Consultative Council on Congenital Abnormalities in the Yarram District*. Canberra.
9. Department of Heatlh (NZ) (1978). *2,4,5-T and human birth defects*. Wellington.
10. EPA (1979). *Report of Assessment of a Field Investigation of Six Year Spontaneous Abortion Rates in Three Oregon Areas in Relation to Forest 2,4,5-T Spray Practices*. Washington: Environmental Protection Agency.
11. Wagner, S. L., Witt, J. M., Norris, L. A., Higgins, J. E., Aresti, A. & Ortiz, M. (1979). *A scientific critique of the EPA Alsea II study and report*. Corvallis: Oregon State University.
12. Crow, K. (1978). Chloracne: the chemical disease, *New Scientist*, 13 April 1978, 78–80.
13. Young, A. L., Calcagni, J. A., Tahlken, C. E. & Tremblay, J. W. (1978). *The toxicology, environmental fate and human risk of herbicide orange and its associated dioxin*. Brooks Air Force Base: USAF Occupational and Environmental Health Laboratory.
14. Lathrop, G. D., Wolfe, W. H., Albanese, R. A. & Moynahan, P. M. (1984). *An Epidemiologic Investigation of Health Effects in Air Force Personnel Following Exposure to Herbicides. Baseline Morbidity Study Results*. Washington: Surgeon General, USAF.

15. Donovan, J. W., Adena, M. A., Rose, G. & Battistutta, D. (1983). *Case-control Study of Congenital Abnormalities and Vietnam Service (Birth Defects Study)*. Canberra: Australian Government Publishing Service.
16. Evatt, P. (1985). *Royal Commission on the Use and Effects of Chemical Agents on Australian Personnel in Vietnam. Final Report*. Canberra: Australian Government Publishing Service.
17. Pearn, J. H. (1983). Teratogens and the male: an analysis with special reference to herbicide exposure. *The Medical Journal of Australia*. July 9, 1983, 16–20.
18. Khera, K. S. & Ruddick, J.A. (1973). Polychlorodibenzo-*p*-dioxin: prenatal effects and the dominant lethal test in Wistar rats. In *Chlorodioxins: Origin and Fate*. Advances in chemistry, series **120**, pp. 70–84. Washington: American Chemical Society.

14

The public's perception of the herbicide problem

I hardly need to present to you examples of the public interest in the issue of 2,4,5-T, dioxin and Agent Orange. The topic has been prominent in the media since the late 1960s. In Table 14.1 I have gathered together some of the headlines taken at random from the Australian press. Similar headlines will also be familiar in the USA, and, to a lesser extent, in the UK.

Firstly, I wish to discuss the herbicides themselves, 2.4-D and 2,4,5-T, with the inevitable dioxin content. Then I will go on to Agent Orange, which has generated some specific issues extra to the 2,4,5-T debate. Dioxin as a chemical in itself will not be discussed here; the issues of waste disposal are apart from the arguments I am considering here, and are raised in Chapter 11.

The health problems reported for the herbicides have largely been those that have some time delay between exposure to the chemical and the realisation of the toxic effect, such as the birth of deformed children to mothers previously exposed. The Vietnam veterans' claims are even better

Table 14.1. *Headlines taken from the Australian press concerning the herbicides 2,4-D, 2,4,5-T and related compounds*[a]

Veterans tell of illness and deformities
New risk found in herbicide
No link yet on sprays – dept study links herbicides with cancer danger
Herbicides complicate suicide case
Scientific studies link cancer to herbicide sprays
Residents hold up herbicide spraying
Checks soon on herbicide levels in babies
Unions will put curbs on suspect herbicides
Herbicide sale continues despite row
New 2,4,5-T warning

[a] The entries are in chronological order, dating from January 1980 to January 1983.

examples of delayed effects. I am sure that this concern arises partly from experiences such as those with asbestos, in which case the delay may be very long. When this is coupled with the natural concerns about childbirth and the very complex feelings associated with malformed children, one can understand how suspicions can arise. The tragic experience of thalidomide is still also in the public mind. The natural rate of birth of children with severe or minor malformations is about 1–2% of live births (Chapter 7). The reasons for the birth of such children are complex, but for the parents of any one particular child there is the need to search for a reason, to offset the irrational feeling of guilt the situation brings. Chemicals are as good a reason as any; better, in fact, as they are less well understood. Why should herbicides in particular be selected as the target for this criticism? I do not know; the selection partly stems from attempts to stop the defoliation programme in Vietnam. Herbicides are very widely used, the spraying is obvious and the effects on vegetation are dramatic, therefore the process is well exhibited to the public. Also, herbicides do not have a monopoly among chemicals as a target for criticism.

Once criticism has been made of a chemical, it is extremely difficult to allay the fears of the public. This is largely because the public does not understand scientific issues; it does not understand the dose–response relationship, nor the difference between 10 ppm and 0.01 ppm of dioxin in 2,4,5-T. The organisations that inform the public very often do not understand the issues either. The media in general do not have journalists with a deep scientific background, and governmental public relations men are often passing on information that they themselves do not fully comprehend.

The newspapers and television like issues that are sensational and alarm the public. They like elaborating on possible dangers, whereas a reasoned argument that puts matters in perspective is not seen as appealing to the public. Particularly good are issues which are complex to the point that categorical statements cannot be made (e.g. 'This chemical is not toxic'). There are then all sorts of opportunities to take statements out of context in these complex matters. The only brake that controls the media is that of advertising revenue. I see plenty of advertisements in the newspapers for cigarettes, none for agricultural herbicides.

Governmental agencies can do an effective job in supplying information to those that ask for it, for they are not in the business of pushing propaganda or force-feeding the public. However, if an issue suddenly becomes prominent, as a question in Parliament or pressure on the Minister, then speed becomes the main consideration in supplying an answer. Such answers may not, therefore, be as scientifically sound as they could be, since the chain of desks through the bureaucracy may be long. The scientist may not have been contacted in time. My own impression of government departments is that they are essentially honest in attempting to supply information, but that the long delays in the bureaucracy give the public the idea that this information is being suppressed. You get the choice between an off-the-cuff answer, which is factually dubious, or a considered reply, which may come next year.

Therefore the public can be excused for being confused about herbicides, or any other technical issue, for that matter. It is just not getting good information.

Another problem that I can see the public has when it does get technical data is that of putting the facts into a reasonable context. What does the factual material of the previous chapter mean? It means that the benefits and disadvantages of the use of 2,4,5-T must be determined on as wide a basis as possible, i.e. the assessment must encompass economic, environmental, human health and other social factors. Exactly the same applies to the use of any other chemical, such as aspirin. If one did an assessment as above for aspirin and for 2,4,5-T in parallel, which would come out in the best light? I would not like to chance a guess. The acute toxicity of aspirin is the same as for 2,4,5-T; chronic dosing can cause gastric ulceration, anaemia, coma due to depression of the central nervous system and cardiovascular collapse, among many other unpleasant problems. Many people have suffered kidney failure and death resulting from the use of mixtures of aspirin, phenacetin and codeine. Aspirin is an effective, useful and relatively safe drug – but not entirely safe. Compare with my comments on 2,4,5-T. Compare the risks of being a herbicide spray operator, with that of coal-mining. In a recent news item, 98 miners were reported dead in a mine disaster in Taiwan. Many more gone in one instant than dioxin has ever claimed. Compare, always compare. The public has to have a yardstick, something to give meaning to quantities and numbers.

You may get the impression from the above that I reject entirely any notion of health problems with herbicides. This is not so, firstly because you should have realised by now that I do not hold such absolute views (as they are contrary to the philosophy of toxicity) and secondly, for an interesting reason the significance of which has not been widely grasped. It is very well for me and other scientists to sit at our desks, review the data on the toxicity of herbicides, then say 'If these chemicals are used as directed for the purposes recommended, then there will be no health problem to the operator or nearby residents'. But I am not so naive as that for I know what happens in practice. The safety directions on the can are not read, the contents are slopped into the sprayer, the wind changes and blows the spray back on the operator, who has not bothered to put on his protective overalls. He then wipes his hands on his wet clothes and sits down on a nearby bank to eat his lunch. Well! Is the chemical safe now? One begins to have doubts. I hasten to add that herbicides have undoubtedly been handled in this manner for years. This casual handling has caused problems with other chemical types of herbicide, but not noticeably so with the 2,4-D and 2,4,5-T family. However, any discussion of the safety of chemicals must consider what actually happens in practice in the field and orchard.

This raises a possible health problem that is not well characterised, but should be examined further. Occasionally reports have been made of a 'neuropathy' resulting from exposure to 2,4-D/2,4,5-T. Early observations on 2,4-D are summarised in reference 1. If this neuropathic syndrome is real and associated with herbicide exposure, which is by no means certain, then it is probably linked to contamination of the skin during handling of the chemicals. It may be a general syndrome relating to organic solvents in general, rather than herbicides in particular. There are plenty of additives other than the active ingredient in a spray mix.

We must pay more attention to how chemicals are handled. The spray operator and the Vietnam veteran may be speaking the truth when they say 'We were soaked in the liquid'. It should not happen, but it does. One inquiry (2) has commented on the need for better practices for the handling of pesticides by (Australian) defence personnel.

It is argued by some scientists and others that we should use different herbicides, since there is a dispute as to the safety of the chlorophenoxyacetic acid class. This has been discussed by Hay (3) in terms of alternative, newer chemicals that have lower toxicity than 2,4,5-T. These chemicals (Roundup, Amicide, Krenite) certainly look promising and the cost of them (now higher than that of 2,4,5-T) will fall as production and sales increase. However, how much do we really know about them? Do we know of the full environmental effects they may exert over periods of 30–40 years? No. Do we know if there are any subtle health threats to spray operators who handle them every working day? No, but we have answers to these questions for 2,4-D and 2,4,5-T. After all the disputes, research, controversy, inquiries and general agony about 2,4,5-T, we know far more about it than we do about any other group of herbicides, or most other domestic or industrial chemicals. Why throw all this away to start again on the unknown? Historically, dioxin has been the main problem with 2,4,5-T; it is now recognised and controlled.

The other alternative sometimes discussed is that of biological control, i.e. find an organism that will kill weeds in sugar cane selectively, or consume unwanted brushwood and blackberry vines. Biological control has its own problems, however, and it is very difficult to find an organism that will obligingly do what we want it to do.

Next we should discuss the conspiracy theory, i.e. that industry and governments are deliberately falsifying, or selectively releasing, information for the purpose of allaying justified public concern about herbicides of the 2,4,5-T type. I cannot believe that scientific data are being falsified on any significant scale, simply because it is too risky. The scientific community is split between laboratories in industry, in government agencies, and in the universities. These three groups are rather jealous of their own standing and watch one another closely. Any consistent production of suspicious results by one group would be jumped on by one of the others. Here the universities would be the independent watchdog (not that government scientists have much love for industry, their job is to watch them). Any collusion by way of business lunches or expensive junkets is usually at a level above that of the working scientist. I can imagine two forms of data falsification that could possibly occur. One can occur at the bench level during routine testing and is known as the 'sharp pencil' technique. The technician cannot be bothered actually doing the test, so merely pencils in an expected result on the report form. The other is deliberate distortion of results at the managerial level to achieve a favourable picture of a product. Both forms of falsification can, however, be exposed by independent checks. If falsification is dangerous, suppression is possible. Industry is not going to divulge unfavourable information about its products unless it is forced to. The job of the government regulatory agency is to squeeze the information out of industry. Sometimes it fails, sometimes the political situation may suggest to the governmental agency that it would be better to lay off industry for a while. Regulations are becoming stricter, and industry now has to supply much more information before it can market a product. The university scientist is again useful as a neutral point of reference. This researcher may well have to duplicate data that industry already has, but this is no waste of resources for confirmation from an independent source is very reassuring, no matter how good you feel the original experiments were.

The presence of a toxic contaminant in 2,4,5-T preparations is reported to have been known to the manufacturers (Dow) in 1950, and to the makers of trichlorophenol in Germany (Boehringer) in the mid-1950s following the discovery of the causative link between dioxin and chloracne (3). This information was not published, but the manufacturers made some attempt to limit the dioxin content of 2,4,5-T. The significance of dioxin as a contaminant caught public attention in the late 1960s when the teratogenicity of 2,4,5-T became known. The accusation is that Dow and other companies suppressed information on the dioxin content and associated toxicity of 2,4,5-T. Perhaps they did, depending on how you define 'suppressed'. No company is going to voluntarily publicise unfavourable information on its products. If the public requires that that sort of data be known, then it must legislate to that end. The defaulting company can then be prosecuted. The lesson from this episode is that good, regulatory legislation is necessary. The industry will moan and groan about such regulations, and may even persuade the public that it is another advance of useless bureaucracy. You, as the public, must know what you want. This topic is discussed further in Chapter 15.

Suppression of information by government agencies is possible. I know of no accusations that this has occurred in the 2,4,5-T story (excepting the general controversies surrounding Agent Orange). Slowness in replying to questions has been mentioned before. Government departments are, of course, subject to political control by the elected representatives of the people. Questions on technical matters (such as the safety of 2,4,5-T) can be addressed repeatedly to governments purely as a political tool in order to embarrass or to disconcert the party in office. For example, I have known the same question to be asked repeatedly (in slightly different form) to the same Minister, on questions that a little research could answer. The public should not mistake the placing of such questions as being founded on genuine concern; it is a political ploy. The large number of official reports from all governments on the chlorophenoxyacetic acid herbicides does not indicate a suppression of information.

So much for the conspiracy theory on 2,4,5-T. What I see is a desire by industry to retain maximal profit, and inexperience in the handling of technical matters by governmental agencies.

This concludes my assessment of the public perception of 2,4,5-T and 2,4-D. What have we learnt from this discussion? The public needs information presented in an accessible form, i.e. related to yardsticks and measures with which they are familiar. Much more attention must be paid to how herbicides (and other chemicals) are actually handled in the field, in contradistinction to what the instructions are. We should be wary about embracing new chemicals in place of established ones, which may have problems, but which are known problems. Lastly we must recognise that industry runs on profit. If we want industry to do something that will cost them money, then it has to be ordered to do it by regulation. We can now move to that most complex of topics, Agent Orange.

The first point I want to raise is one which has not figured greatly in the public debate, but which I personally see as important. That is that the use of Agent Orange, whatever its human cost, was a great blow to the environment. I feel we should be kind to the environment as it is the legacy we will pass to our heirs. Throughout history people have been bashing, butchering and abusing each other, but the results do heal, in a long timescale. Genghis Khan and Timur Leng did a good job of population control in Asia, and the Thirty Years War depopulated Germany for 100 years, but eventually the wounds healed and people flourished again. Our environment is now, however, in a condition that is very delicately balanced; any serious assault on it may cause irreversible damage. Part of the opposition to the defoliation programme in Vietnam did arise from concern about the ecological damage that might be done, but that concern was from academics and it did not become a popular cause. Orians & Pfeiffer (4) described the short-term results of defoliation in Vietnam; they also mention the equally devastating effects on the environment of bombing with 500 and 750 pound bombs. I do not know whether the ecology of the country has now recovered; a report from an independent source is not available. Ecology has been an issue for academics: human health has dominated public discussion.

There are two groups of people whose health could have been affected by Agent Orange. These are the inhabitants of the country (civil and military) and the military from outside who fought there (Americans, Australians, Koreans, etc.). To discuss the health of the former group is difficult as we outside the country cannot get direct access to information, and may suspect that what we are told is biassed. Therefore I will confine this discussion to the health of US and Australian veterans. There are three questions to be answered.

1) Were the personnel involved actually exposed to Agent Orange?
2) Do they show effects on health not shown by similar people in similar conditions, who were not exposed to Agent Orange?
3) Can these adverse effects be shown to arise from the exposure to Agent Orange?

The public and media seem to have the attitude that everyone who served in Vietnam was exposed to Agent Orange. This seems unlikely, to put it mildly. One feature of the war was the high ratio of the support troops to the men engaging the enemy in the field. Some support units were undoubtedly in sprayed areas. However, how many men served their time in Saigon and never went into the forests? The question is more complex for Australian troops, as Agent Orange was not an Australian military store. Any exposure would be as a result of a joint exercise with US troops, or by contamination resulting from some mistake. An attempt has been made to estimate what degree of exposure Australian troops did receive to US herbicides (2). There was one known occasion when they were directly exposed to herbicides from a Ranch Hand mission. The estimate of what other exposure occurred develops into arguments about spray drift and deviations of US aircraft from flight paths. One estimate of the possibility of exposure was 5% of the Australian force; the Vietnam Veterans Association of Australia was of the opinion that all were exposed. If we accept chloracne as a good indicator of exposure to dioxin, then very few veterans were exposed to that chemical. Chloracne is not merely a skin rash, which is fairly frequently reported. From a scientific point of view, the question of exposure cannot be ignored. Some attempt must be made to gauge what degree of exposure the veterans received, even if this is done only by military unit rather than by individual.

When we come to the second question, the health of veterans, we enter another uncharted sea of hazards. There is undoubtedly a Vietnam veterans' syndrome. Whether it is wholly psychological, or due to exposure to Agent Orange, or to antimalarial drugs, or to marihuana or whatever, we do not know. In this situation the proper scientific course is to characterise and define the syndrome, then examine each of the possible causative factors with equal thoroughness, until the factor or factors responsible are found. Strangely enough I do not remember seeing a broad study of veterans' health among all the reports and studies so far made. By broad I mean one that surveys the health in a wide vista, including physical health, psychological health, social problems, attitudes to the community, use of alcohol, etc. The veterans' insistence on Agent Orange as the cause of their problems has done themselves a disservice, as all the inquiries go over the same old fruitless ground of supposed birth defects.

My own impression is that the veterans are just cheesed off with the Vietnam episode and have seized on one concrete issue as the scapegoat, rather than the whole political–social complex which is their real complaint. These men were required by their countries to fight a war. They did that to the best of their abilities and many suffered for it. On their return, they were not heroes; a large part of the home population just wanted to forget them. Whatever the rights or wrongs of the Vietnam war, the veterans did the job they were given. Political decisions are made by governments; the consequences should not be held against the servants of those governments.

I remember the attitude of my contemporaries in 1956 during the Suez crisis in the UK. Many of these young friends had just completed 2 years of National Service. During the crisis, those on the Z Reserve were recalled for 6 months or so in camps in the UK. This further interruption to their jobs, studies and careers engendered a bitter resentment against the 'system', even though these men were patriotic in their attitudes. There was talk of near-mutinies in some of the camps. I can sense something of the same feeling among the Vietnam veterans.

We cannot confine a discussion of the veterans' health to Agent Orange. Considering purely chemical factors, there are many that could conceivably have affected them (that is, if they have a chemically induced problem at all). Insect repellents, insecticides, antimalarial drugs, explosives and propellants were used, with the social drugs of alcohol, marihuana and possibly others (see Table 7.5). Initial psychological problems of readjustment are likely to have been compounded by alcohol, which has a special place in the military society. Every mess is a drinking place where one learns the rituals of the alcohol ceremony. This is carried on into the clubs for veterans at home. Therefore the effects of this alcohol culture must be reflected in the veterans' health.

Can these ill-defined health effects be shown to have been caused by exposure to Agent Orange? No. At the moment the health problems need definition before any cause can be found. The specific claims of the veterans' relative to Agent Orange have not been substantiated, as seen in the previous chapter.

Once again we ask why Agent Orange has become the centre of this problem. One reason is, I believe, more or less pure chance. The public had an unease about chemicals in general, and this crystallised around 2,4,5-T. The veterans sought a reason for their problems and subconsciously chose Orange. Another reason is political. The Vietnam War was fought and won as much on public opinion and politics as on military capability. The defoliation programme was a splendid propaganda opportunity for the North Vietnamese, which they did not neglect. The Soviets and North Vietnamese still push this issue to the point that public opinion now readily equates the use of herbicides to chemical warfare. This equation is nonsensical in scientific terms; the toxicity of chemical warfare agents (Table 4.2 and 10.1) is 10 000 times that of 2,4,5-T. You cannot compare a chemical made specifically to kill humans with a chemical that has been used in agriculture for 40 years by the thousands of tons and not shown to have any marked human health problem. Or you can make such a comparison, if your aim is to deceive the public for reasons of your own. The propaganda campaign about Agent Orange has been surprisingly successful, as I have found from my own experience. I recently spent some time describing to a senior government official the protective capabilities of the chemical suit now available to the Australian Army. After I had described the protection afforded against nerve agents and mustard

(Chapter 10), his comment was: 'Will it protect against Agent Orange?'; so well entrenched is the concern about this herbicide. I stress that I do not see political propaganda as the only cause of the public's concern on this issue; it may be a minor cause, but effective.

Undoubtedly one of the major problems in the whole herbicide issue and the public debate is the refusal of many parties to look at the facts. This is not just a failing of the public, which may be excused on the grounds that it is ill-informed, but also of many academics and others to whom the facts are accessible if they care to seek them. Dr Jock McCulloch (5) sees many of the issues discussed above concerning Agent Orange with extraordinary clarity and simplicity. According to him all US and Australian troops were exposed to pesticides and the symptoms exhibited by the veterans are entirely consistent with the known toxic effects of such chemicals. The pesticides he cites include chlorophenoxyacetic acid herbicides, organophosphate insecticides and chlorinated hydrocarbon insecticides, each of which class has quite different toxicological properties. Has Dr McCulloch linked these properties with the (ill-defined) symptoms of the veterans? Not to my knowledge. Dr McCulloch has failed to establish the factual basis of his arguments at source, and thus built his subsequent conclusions on air. Thus his book (6), whatever its merits as a socio-political discussion, assumes what is scientifically very uncertain. Dr McCulloch has started from a premise of certainty; reason must allow an element of uncertainty, in this case a great element.

Others have looked carefully at the facts, but produced arguments that are so tenuous that it is hard to see that they add anything of value to the debate. An example is the hypothesis of Drs Hall and Selinger that excess Agent Orange from the Vietnam War was sold to Australia through Singapore, that this herbicide was used in Australia, and could have caused increased birth defects, as in the Yarram district of Victoria, which incident was mentioned in the previous chapter. Their arguments appear in a number of articles (7a & b). The causal chain of events claimed by these authors is given in Table 14.2. As can be seen by the notes, there is not one single point in this argument that can be confirmed; in fact, an inquiry has done the opposite for the Yarram claim. Thus there is not one solid link in the chain. It is unlikely that Agent Orange was sold off by the US Air Force before 1970 as, in fact, they were virtually grabbing all they could get. Attempts to sell off the chemical after the final termination of the defoliation programme in February 1971 were aborted (3) and the stocks were finally incinerated at sea in 1977. Hall & Selinger may possibly be correct that some was sold off to the international market and came to Australia, but their evidence is weak. If the herbicide was imported, the dioxin content on average must be presumed to be below 2 ppm (p. 175). Once in Australia, the material loses its identity among the herbicides from established sources. There is nothing to link the use of the presumed herbicide to Yarram or any other specific place in Australia. Finally, there is no certain indication that the spraying of 2,4,5-T has caused birth defects anywhere in Australia.

The hypothesis perhaps deserves examination, but does it merit publication when the facts are so poorly established and the links between them presumptive only? Read the references for yourself and make your own judgement. The danger is that such hypotheses can turn into circular arguments. There is nothing to connect the presumed import of Agent Orange with a possible increase in birth defects except the prior premise that one is the cause of the other. You cannot then use an observed coincidence in time between the two events to confirm that one is indeed the cause of the other. However, I am willing to believe that this hypothesis will be used as ammunition in support of the Agent Orange birth defects argument.

A note in *New Scientist* (8) points out the dangers of deducing cause and effect from events following one another in time. It points out the perfect relationship in time between two graphs published in another journal, one showing the '$ trade-weighted exchange rate index' and the other the new cases of acquired immune deficiency syndrome (AIDS) victims in the US. The link would statistically appear very sound, and monetary policy proven to be a cause of AIDS, except that the two phenomena are

Table 14.2. *Elements in the presumed causal chain that links excess Agent Orange from the Vietnam War with abnormalities at birth in Australia*

Herbicide in store after spraying suspended in 1970
Sale of excess store on international markets
[Not substantiated]
Import to Australia from Singapore
[Not identified, but presumed to be 2,4-D and 2,4,5-T under a code number for trade purposes]
Usage in Australia as herbicide
[Locations not known]
Exposure of pregnant women
[Exposure seems to be presumed if any spraying occurred in neighbourhood]
Increase in number of defects at birth
[Never been confirmed: Hall & Selinger have some evidence in the Yarram case which can be disputed]

The Hall-Selinger hypothesis, reference 7.

unrelated. This is in fact the argument used by the tobacco industry lobby; the link between smoking and lung cancer is only a statistical one. To a large extent it is, but the difference from the AIDS example is that very many facets of the smoking link have been explored and found to be related to tobacco consumption. These include change in the smoking habits of the sexes, the different tobacco consumption in countries with different cancer morbidities, and change in disease patterns (after a suitable interval) following the reduction in smoking or the reduction in tobacco tar. No such complex pattern of confirmatory evidence exists for a relationship between exposure to Agent Orange and health problems, nor for a link between monetary policy and AIDS.

The scientists and academics have a responsibility to ensure that any argument brought into the public arena for debate has a sound factual background. Hypotheses can be debated in academic or technical circles and, if found reasonable, can then be made the subject of public comment. The problem with a highly controversial topic is that any pronouncement becomes of great interest, and I understand the dilemma of Drs Hall and Selinger, in that any tentative hypothesis is immediately seized on as fact and used to advance one side of the debate. They have made a number of quite proper reservations about their hypothesis, but unfortunately the media and the public sweep aside such reservations; they like simple statements. The consequence is that much more is imputed to these scientists and in much stronger terms than they themselves may initially have wished.

I have just mentioned the presumption that Agent Orange is the cause of birth defects which is central to the Hall–Selinger hypothesis (Table 14.2). It is worth referring to the 'crooked-calf disease' incident mentioned in Chapter 16. The mother initially believed her deformed child was the result of 2,4,-D spraying, but subsequent investigation established anagyrine as a much more likely cause. In a case such as the Yarram situation, if there is indeed a real effect, you cannot take just one presumed cause and argue for and against that. You must take all reasonable causative agents and test each one for candidature as the actual cause. This widely increased perspective may change your thinking dramatically. There are many, many, well-substantiated causes of birth defects; Agent Orange is not one of them.

If we briefly digress to consider what evidence there is to associate Agent Orange with birth defects among the Vietnamese population, then again there is very little. A symposium in Ho Chi Minh City in January 1983 did not produce definite evidence for such an association. It was concluded that there was a need for further study (9), but hard evidence is still lacking.

It is time to look back on our discussion on Agent Orange and see where we have got to. There is a need for comprehensive studies on the health and wellbeing of Vietnam veterans, uncoupled from the current preoccupation with birth defects. This should lead to an understanding of the veterans' problems as a whole, in social and psychological terms as well as physical health. We have to determine who was exposed to which chemical, then try and see whether there are any causal links between such exposure and any defined problem. Account must be taken of pressures such as political and financial ones (the veterans are seeking a financial settlement). We must understand how Agent Orange has become the focal point or battle standard of an issue that really has little to do with the herbicide. Then we have to emphasise the necessity of keeping strictly to facts which can be established and accepted.

The overall conclusion from these two chapters is that public debate on a technical question is hampered by poor information given to the public, which is left to flounder in a maze of technical jargon, half-truths and gross simplification. My purpose in writing the chapters was not to judge and pronounce on the herbicide dispute, but to explore the problems of technical information and discussion, and to illustrate by a real example how public attitudes towards chemicals are formed. Who said chemistry was a dull subject?

We can look into the herbicide dispute and see in it, reflected as in a mirror, all our own attributes: love of controversy, indecisiveness, inattention to facts, rebelliousness, factionalism, self interest, need for reassurance, fear of the unknown. We are in a Chemical Age, but our personal attributes are the same as they ever were.

References

1. House, W. B., Goodson, L. H., Cadberry, H. M. & Dockter, K. W. (1967). *Assessment of Ecological Effects of Extensive or Reported Use of Herbicides.* Prepared by Midwest Research Institute, Kansas City, for US Dept. of Defense. US Dept of Commerce reference AD 824314.
2. Senate (1982). *Pesticides and the health of Australian Vietnam Veterans: First Report.* Senate Standing Committee on Science and the Environment. Canberra: Australian Government Publishing Service.
3. Hay, A. (1983). *The Chemical Scythe: Lessons of 2,4,5-T and Dioxin.* New York: Plenum Press.
4. Orians, G. H. & Pfeiffer, E. W. (1970). Ecological effects of the war in Vietnam. *Science*, **168**, 544–54.
5. McCulloch, J. (1984). Orange: science and politics. *The Age* (Melbourne newspaper), 23 May 1984.
6. McCulloch, J. (1984). *The Politics of Agent Orange.* Melbourne: Heinemann.

7a. Hall, P. & Selinger, B. (1980). Australian infant mortality from congenital abnormalities of the central nervous system: a significant increase in time. *Chemistry in Australia*, **47**, 420–2.
7b. Hall, P. & Selinger, B. (1981). Australian herbicide usage and congenital abnormalities. *Chemistry in Australia*, **48**, 131–2.
8. Anonymous (1985). Note in column headed 'Feedback'. *New Scientist*, 7 March 1985, p. 48.
9. Westing, A. H. (Ed.) (1984). *Herbicides in War. The Long-term Ecological and Human Consequences*. SIPRI. London: Taylor & Francis.

15

The technical contribution to regulatory and legal problems concerning chemicals

On Monday, 7 May 1984, announcement was made of an out-of-court settlement in a class action of Vietnam War veterans against companies which had made Agent Orange for the US Government. Was the $180 million dollars settlement a recognition that the companies were at fault? No, the companies stated that the settlement was not an admission of guilt; it is indeed very unlikely that they had done anything illegal or unethical in the context of the marketplace (Chapters 13 and 14). Was the settlement a sell-out by the veterans' lawyers, who ought to have stood out for much more? Yes, said the veterans' organisations (or some of them). What was the reason for the settlement? Basically it was because there are 600 000 lawyers in the USA.

The settlement was foreshadowed in an address by Paul F. Oreffice, the president of the Dow Chemical Company, to the Society of Chemistry and Industry meeting in New York on 5 October 1983. Oreffice's address (1) concerned the inhibiting effects on industry of the pressure of lawsuits brought against companies for reasons often trivial. Lawsuits may be initiated by a plaintiff through a lawyer willing to work on a contingency basis, that is, if the suit fails, the plaintiff pays nothing. If the suit succeeds the plaintiff gets compensation, minus a large deduction for the lawyer. This leads to the situation where the lawyer is looking for plaintiffs, rather than the plaintiff trying to find, and pay for, a lawyer. Trivial and frivolous lawsuits thus multiply; the plaintiff cannot lose. It is often cheaper for the company to settle out of court, even if the case has little substance.

In British law as current in the UK, Australia and Canada, the situation is different. A party that sues on a frivolous basis and loses, may have the costs of both sides awarded against it. This makes the plaintiff think carefully before going to court, and in these countries the litigation is much less. Oreffice states that this situation in the USA is inhibiting the research and development work that chemical companies might otherwise do. For example, the Dow group has ceased all research for products in the obstetrics area, a difficult area that has attracted public attention because of the risks of harm to the foetus, witness the thalidomide tragedy. British law may reduce the volume of lawsuits, but chemical companies must still guard very carefully against expensive claims for compensation.

To consider the other point of view, the chemical worker and the consumer have suffered from the negligence of industry. There are two classes of incidents here; those in which a specific population of workers or consumers have suffered an obvious disease caused by a chemical, and those in which a wide group of people have suffered very minor but chronic ill health from exposure to chemicals. Examples of the first group (call it class A) are the lung cancers and mesotheliomas suffered by asbestos workers, the bladder cancers caused by 2-naphthylamine, leukemias caused by benzene, and the birth deformities resulting from the use of thalidomide. The cause and effect relationship for these chemicals and diseases was worked out from epidemiological evidence, which was fairly easy as each disease is not common and it occurred among a specific group of people (Table 13.1). These events are tragic, but the dramatic nature of the cases has in itself made the cause detectable. Class B events are much more difficult to detect. An example would be of exposure to fumes from dry-cleaning fluids or degreasing solvents which give the operators headaches and a feeling of poor health over a long period of time, but lead to no obvious disease. A second example would be the exposure of urban populations to two products of car exhausts, carbon monoxide and lead. Each of these two chemicals can easily be shown to be toxic at high dose; the current debate is about whether the chronic low dosing of city dwellers actually affects their health.

The class A chemicals still present legal or regulatory problems, even though the cause and effect relationship is unarguable. However, class B covers all the problems one could ever want; even the 600 000 lawyers in the USA must be happy about this situation.

In this chapter I want to discuss the legal and regulatory aspects of chemicals from the technical point of view, that is, from the contribution that the scientist can make. The structure and formalities of the law in a particular country obviously affect the chemical industry quite markedly, and hence the degree to which society either benefits from, or suffers from, chemicals in the world. Laws and regulations vary from country to country. In fact in countries with a federal structure (USA, Australia, etc.) they can vary from state to state. Because of this complexity of regulations I do not intend to review them in detail. Also I am not a lawyer, so that this review is written from the technical viewpoint, not from that of a legislator, lawyer or consultant.

Chemicals pass from production to use to disposal, so we can consider regulation in these three subdivisions. Industry is involved in the first and to some degree the last. The public is mainly concerned with use and disposal.

The purpose of industry is to make a profit, which is used to reward shareholders in order to maintain investment, to provide for new plant and to fund technical research. It is not a benevolent institution, individual companies survive by being competitive. Therefore if the community wants protection from industry, it must gain it by clear and fair legislation. An individual company will not put in expensive equipment to reduce chemical risks to its employees, if competitive companies are not doing it. The manager is not being inhuman in making this refusal, it is a realistic decision. Money spent on the installation of safety equipment has to be recouped by increasing product prices, which will mean the loss of sales to the competing (less conscientious) firms. The consequence is that the manager may be sacked for weakness. The next manager will be selected to be much tougher. On the other hand, if legislation requires all companies to install the equipment, the managers will again grumble in public but will really be much happier in private, and the job will be done. The problem now is the extent of the regulatory jurisdiction; can the products be brought in over state lines or from overseas locations where regulations are lax? As a second example of this situation, we can take the health foods or synthetic vitamins market, which is now big business (Chapter 9). Controls vary from nil to moderate in various countries. Since the market is competitive, the claims for the products become a little more extravagant each year. The vitamin content is increased a little more, or another natural oil is added, to get ahead of the competition. The conscientious marketing manager may feel that he is producing a product which is becoming questionable in safety to the consumer, yet he is impelled to continue. He may welcome reasonable legislation that controls this situation. The less scrupulous operator is going to protest very loudly.

The public therefore has to see industry for what it is, and think in terms of obligatory controls. Industry must accept regulation, indeed it may welcome it, if the regulation is sensible. Whether it is sensible or not depends on its technical content, and the interpretation of that content by the public and the law makers. Herein lies the difficulty.

Let us examine a simple case in which we want to set a limit to the exposure of workers to a chemical, say trichloroethylene. We wish to regulate exposure to this chemical because we know (2) that high concentrations of the vapour cause temporary unconsciousness and sometimes death. Chronic exposure leads to damage to the heart, cranial nerves and liver. We therefore wish to establish a maximum acceptable concentration of trichloroethylene vapour, below which we believe that a worker will not be subjected to harmful effects (Chapter 4). The mechanisms for setting these hygienic standards varies from country to country. The standards may be suggested by a committee of a learned society involved in industrial toxicology. The Department of Health may set up an advisory committee and appoint consultants with experience in industrial medicine and toxicology. In Australia we have 'hygienic standards' for levels which were derived from Threshold Limit Values (TLVs) recommended by the American Conference of Governmental Industrial Hygienists and then endorsed by the National Health and Medical Research Council (Australia). The level of trichloroethylene vapour in air is 100 ppm. The initial choice of such a level is purely an educated guess, based on information from past exposures. This information is usually sparse and incomplete, for if there are clear cases of ill health resulting from exposure, the doses received by the individuals involved must often be estimated roughly. One has to guess and then allow a generous safety margin (say 100-fold). Once the level has been chosen, it is up for discussion and modification if this is justified by further experience (Fig. 15.1). There is just no way of making the process more exact; the recommended level must be understood as a crude estimate only. Thus there is no difference at all between 90 and 110 ppm of trichloroethylene, we certainly cannot say that the first concentration is safe and the second is dangerous. We should aim to keep the concentration as low as possible at all times, and ensure that the time-weighted average concentration over an 8 hour working day never approaches 100 ppm.

Once a limit of this kind has been set, it may have only advisory force, or it may be incorporated into obligatory regulations. It is preferable if the latter is the case, however a manager would be foolish to ignore an advisory level, as any lawsuit for damages would have much greater chances of success if it could be proved that an advisory level had been exceeded.

This all sounds nice and straightforward. Or is it? The data on which we rely to set recommended levels are so sparse that they can be interpreted and extrapolated in many ways; different people can draw quite different conclusions from them. This has been discussed by Elkins (3) on the basis of many years experience. He points out that seven cases of leukaemia caused by benzene, in a single company over a period of 20–30 years, were used to predict 1400 deaths by benzene-induced leukaemia per year (in the USA). The concentrations to which the workers were exposed were unusually high, but unknown. Thus the estimation of 1400 deaths per year results on the extrapolation from one death every 3 or 4 years, dosage unknown. There is a whole history of other effects due to benzene, all anecdotal and most without good concentration data. These are the sort of data that have to be used to set maximum levels, supplemented by information derived from animal experiments, again difficult to extrapolate to humans (Chapter 4).

Apart from the paucity of information, its interpretation is subject to argument, particularly the relationship between response and dose. As this is a subject of much current debate, both in respect to chemicals and to radiation, it has been discussed fully in Chapter 4. My conclusion from that discussion is that no dose is absolutely safe, but that very small doses present no practical risk. An arbitrary decision has to be made as to what level of chemical can be considered safe in all likely circumstances.

Fig. 15.1 An exposure chamber in which volunteers can be exposed to concentrations of solvent vapours up to permitted industrial levels. The person at left is supervising the instruments which measure and control the concentration of solvent vapour in the chamber. The psychological testing procedure is being supervised by the person seated at right. Such tests allow further insight as to the adequacy of current permitted levels, and give data to support the development of better methods to monitor the uptake and clearance of industrial solvents (Crown copyright, from the 1980 issue of *Health and Safety Research*, produced by the UK Health and Safety Executive).

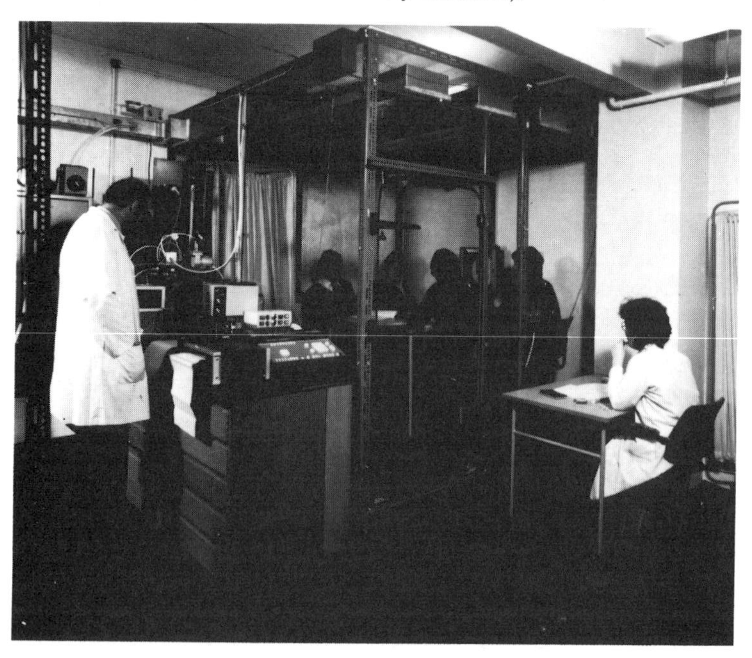

Someone is going to say now: 'Why not zero level?'. Because as we have seen in the chapter on analysis, zero levels do not exist. If we keep on improving our methods of analysis we will find trichloroethylene in parts of the factory we previously thought clear of it. We will find trichloroethylene in the air if someone takes the stopper out of a bottle two rooms away.

Industry as a whole tends to follow established work patterns, once a process has been set up and commissioned. The chemical research laboratory is different, because new chemicals of unknown toxicity are being made and handled in it. Therefore we cannot rely on regulations for specific chemicals to control exposure, otherwise the research chemists could do nothing. Here the wisest course is to limit the exposure to all chemicals to the minimum possible, by means of the best equipment and good handling procedures. To an extent we can rely on the chemist's own discretion to look after himself, but not entirely. Historically, the chemical synthesis laboratory was the messiest of them all. I used to be very unwilling to lend equipment to the synthetic chemists, as it would come back corroded and spattered with unknown material. Nowadays, they are more conscious of chemical hygiene.

We will leave industry for the moment, and have a look at chemicals in use, a topic which very largely affects the general public.

It is hoped that the grossly toxic chemicals are recognised and banned before they are released to the public. If the toxic effects are subtle, then the problem is more difficult. Such an effect can be a long-delayed response, as to asbestos inhalation, or an effect in a special situation such as the use of thalidomide in pregnancy. These make a challenge to the regulatory authorities, who have to insist on more complex tests to screen out these subtle effects. Practically all the hazardous chemicals in the public domain belong to my class B; the only non-industrial class A one I can think of is thalidomide. This does not make it any easier, in fact it is much harder to recognise and deal with class B compounds.

We are at present moving from an era of what we might call retrospective regulation to one of anticipatory regulation. In the past, chemicals were made and used freely. If there was a problem, then it was eventually recognised, and slowly some control would be introduced (extraordinarily slowly in some cases – consider asbestos or 2-naphthylamine; Chapter 8). Anticipatory regulation should be much better; we would examine each chemical before it went to the public. There are problems, however, many of them. Firstly, there is the enormous number of chemicals already available to the public, many of which have not been characterised at all as regards toxicity. Secondly, where do we stop? Horseradish sauce is a very complex mix of chemicals, practically none of which has been characterised. Do we ban sales of horseradish sauce for 10 years whilst we do toxicity tests? If we exclude traditional foodstuffs from our screening (rather rashly, as we contact them most intimately) then what about other natural products? Evening primose oil is now a fashionable dietary supplement along with Omega-3 fatty acids from fish oil. If we test all similar products, we are setting an impossible task. The natural health enthusiast will insist that we do not need to test 'natural' chemicals because, being natural, they are naturally good for us. This is pure rubbish, as we have seen in Chapter 9, there is no distinction between natural and synthetic chemicals (except in associated impurities), and there are plenty of positively harmful natural products (e.g. the carcinogens discussed in Chapters 8 and 16).

All that can be done by anticipatory regulation is to make intelligent guesses about the possible toxicity of synthetic chemicals, then subject them to appropriate tests before they are cleared for release to the public. We can also rank the novel chemicals in an order which expresses the requirement for strictness of testing. In order to make the testing appropriate, we must also make an intelligent assessment of the type of test necessary for the particular chemical; this is what I have summarised under the heading 'Screen' in Table 15.1. Here are shown in outline the type of questions we need to ask about the candidate chemical, and the testing necessary. We are most concerned about chemicals that will contact us (or be in our food) for long periods and therefore have the longest time to exert an effect. Structural chemicals do not worry us much, unless a small molecule contaminates the polymer and can diffuse out. There has been recent concern about the evolution of formaldehyde from urea-formaldehyde resins used for structural purposes, including insulation.

Given that we need to test chemicals for toxicity at doses appropriate to the possible exposure of humans, how good are our tests? The answer is probably that they are dependent on the thoroughness with which they are pursued, which again depends on the availability of time and money, and on the intelligence of the researcher. Test methods are discussed in Chapter 4 and summarised in Table 4.3. Acute lethality tests are quite simple. As one often requires an order-of-magnitude answer only, you can dispense with the mass slaughter of animals required to get an LD_{50} value accurate to the second digit. The chronic forms of testing are much more costly and difficult. The Ames test for mutagenicity and other short-term tests (4) are attempts to develop rapid screening tests that do not involve experiments on living animals. The Ames test can give useful information, but cannot be regarded as definitive. The test, as a screen for mutagenicity, does not always correlate well with carcinogenicity (4), but it (and similar short-term tests) are valuable if assessed in the right way (5). We may

Table 15.1. *Outline of toxicity testing requirements for the classes of synthetic chemicals*

Class of chemical	Screen	Toxicity tests[a]
Structural	Check for remaining monomers or low molecular weight contaminants	Chronic toxicity of any found at appropriate concentration levels
	Will use involve prolonged contact with skin?	Check for skin irritation, allergic dermatitis, etc.
Pesticides	Estimate degree of exposure of spray operators in realistic conditions of use	Acute and chronic toxicity
	Check if residues will appear on foodstuffs	Test chronic oral toxicity at relevant concentrations
	Environmentally stable	Check toxicity to environmental organisms and humans at appropriate concentrations
	Not stable in environment	Check toxicities of breakdown products
Drugs	Short-term use for acute medical conditions	Acute toxicity, some attention to chronic effects
	Use for long-term to control a chronic medical condition	Special attention to chronic toxicity
	Potential for abuse by public	Appropriate tests depending on form of abuse
Process	Food additives, or could contaminate food	Test thoroughly for acute and chronic effects
	High degree of contact with humans (mainly domestic use)	Test acute and chronic effects
	Low degree of contact with humans	Test as appropriate
	Environmentally stable	Test as appropriate to means of disposal and likely fate in environment

[a] Toxicity includes all adverse effects such as carcinogenicity, teratogenicity etc.

conclude that we have reasonable technical ability to conduct tests on the safety of chemicals, but this is an extremely costly and slow business.

It has been suggested that since there is an enormous amount of knowledge available about the structure of chemicals and their toxicity, it should be possible to make deductions about the relationship between the two properties, and therefore predict the toxicity of a new chemical from its structure. True, this topic has been studied for a long time, and many reports have been made on structure–activity relationships for various classes of chemicals. Unfortunately our knowledge of why a chemical shows biological activity is still at a stage where we cannot confidently make predictions. For example, the organophosphate esters have been studied for a long time. We could make some intelligent guesses about the toxicity of a new member of the class; for example, if it had no obvious leaving group, we would be reasonably confident it was not an inhibitor of cholinesterase. We would be less confident about its neurotoxic properties, as we know less about this subject. Interference with the metabolism of natural esters of phosphorus (water and fat soluble) could be gauged from its structure. However, we could not guarantee the absence of some totally unexpected manifestation of toxicity, dependent for instance on a physical property such as the water/lipid partition ratio. The antivivisectionists have raised this argument as showing the unnecessary nature of animal testing; it can all be done on a computer. Unfortunately it cannot. I have some sympathy with the antivivisectionists' arguments. There is a case against slaughter of animals to test what can be deemed an unnecessary chemical, for instance a cosmetic, but such tests may be necessary if the chemical has obvious human benefits. I am afraid we are stuck with animal tests for a long time ahead.

Anticipatory regulation of chemicals thus brings enormous practical problems. We may hope that novel chemicals introduced for use by the public will be tested and relatively safe. The enormous number of chemicals already in use will continue to be screened mainly by the retrospective method, with all its potentialities for delayed tragedies.

We have looked at production (industry) and use (the public), let us now briefly consider disposal, which concerns both industry and the public. In former times it seems that disposal was never considered. When the chemical was finished with, you flushed it down the drain or carted it to the tip, and said goodbye to it. Unfortunately, it did not in many cases just silently disappear into the environment, and as a result of many nasty experiences with fluoroacetate, methylmercury, dioxin and so forth, we cannot any longer use our naive dumping techniques. Considering the vast quantities of chemicals that are produced, it is astonishing to me that their disposal has been so little studied. I don't mean just dramatically toxic chemicals such as dioxin, but all of them. Their sheer bulk makes disposal a topic, even if health problems are non-existent. We can require that when a new chemical is introduced to the market the manufacturer has to give some indication of its eventual fate. Biodeterioration is certainly a useful idea. In fact the types of deterioration that are regarded as disadvantages in structural chemicals might be positive advantages as far as the environment is concerned. Perhaps we should increase the sensitivity of polyethylene to ultraviolet light so that all those plastic bags that blow around everywhere would disappear faster. My fear is that we will just disappear among a pile of chemical junk; I do not think toxicity is our main problem.

As matters stand now, toxicity is the main concern as regards chemical disposal. In technical terms we can tell the legislators what we know about chemicals and the environment (Chapter 11) and from this partial knowledge we must construct reasonable rules and regulations for the disposal of wastes.

It is not wise to ban completely the disposal of industrial wastes, then to walk away and forget the problem. This would just invite much more illicit disposal. All chemical companies and research laboratories have stores of the more toxic wastes they want to get rid of, and are holding pending clarification of legislation and the means of disposing of them. I am referring to known carcinogens such as benzidine and 2-naphthylamine, and any highly toxic chemicals resulting from research work. We require the technical means to destroy totally these chemicals (at present high-temperature incineration seems to be the answer) and the consent of the regulatory authorities for it to be done.

What can we offer as technical advice to the legal system to help resolve chemically related litigation? There are three groups whose arguments will be heard by the lawyers.

1) The industrial companies, technically competent and able to present a polished case, but of course with a direct interest.
2) The public, represented by varied groups and associations, with a widely varying technical competence ranging from incoherent crank ideas to the best university science, mainly very sincere in their intentions.
3) The regulatory authorities and their technical advisers, competent but constrained by their position as a buffer between the first two groups and by political considerations imposed by their masters of the moment.

All the arguments are presented by and adjudged upon by persons with legal training, not by scientists. We must therefore endeavour to make the lawyers understand the basic arguments. We would be much more successful in this aim if the scientists did not spend their time arguing about these basics, however there is a reasonable consensus to support my conclusions. We have discussed these elsewhere in this book, so I have collected the major topics and the principal conclusions in Table 15.2, with references to the appropriate chapters. I accept that some of these conclusions are still open for discussion. A nice instance is that of a symposium in 1975, in which one speaker expressed the view that we should press for zero exposure to carcinogens, and the following speaker pointed out the practical difficulty of a zero-exposure concept. Epstein (6) stated that the concept of a threshold level for a carcinogen has no practical significance because we have no method of establishing such a level, therefore we should prevent any exposure to a carcinogen as detectable by the most sensitive method available. The arguments advanced by Westerholm (7) against a zero-exposure concept are essentially those I have given elsewhere (Chapter 6). I believe Epstein is largely correct; as we have seen, the establishment of threshold levels is a very uncertain, arbitrary procedure. However, I think he is solving one problem by substituting another one. To use the most sensitive detection method

Table 15.2. *General conclusions relating to the regulation of risks from chemicals*

Subject	Conclusions
Dose–response relationship	1. No chemical safe at all levels. Chapter 4.
	2. All chemicals toxic at some dose levels. Uncertain what happens at lower dose levels.
Allowable concentrations or threshold levels	3. A matter of relative risk. No entirely safe level (consequence of item 2). Chapters 4 and 7.
Zero allowable level	4. Technically meaningless. Zero level is only a measure of your technical capability to detect a chemical. Chapter 6.
Causation	5. Assignment of cause can only be expressed as a statistical probability. This is a consequence of the origin of data from epidemiological surveys and laboratory experiments. Chapters 7 and 14.
	6. Any toxic effect is almost always due to a number of separate factors. For example, Cancers (Chapter 8).

available means that the regulations are shifting and uncertain from year to year. What was judged completely safe last year now becomes dangerous, not because the working environment has changed, but because we have made an advance in the techniques of detection. I believe we have to accept and live with the two facts: that threshold levels are largely arbitrary, and that zero levels do not exist.

I can also see legal problems arising from the topic of causation. Lawyers are trained in a system of guilty or not guilty, a yes or no result. How are they going to cope with the increasing number of cases that are not yes/no, but depend on probabilities, and on multiple cause. Of course they are used to assigning blame among parties and apportioning damages accordingly, but they do not seem too keen on this. Motor accident cases here are usually settled in terms of blame on one party only; very rarely is it conceded that each party has contributed to the accident.

'Are you sure that chemical *X* caused this disease?'

'Our epidemiological studies show that 90% of the cases in this industry were workers who had handled this chemical. Our client handled this chemical'

'Can you be positive that in this particular case *X* is the cause of your client's disability?'

'No. But the connection is strong'

'However, 10% of the workers who have had the disease never handled *X*?'

And so on.

Why not consider a simple, clear-cut case. A worker who spent 5 years handling blue asbestos in a factory 20 years ago has just died of lung cancer. His wife is suing the factory for damages. No one can argue this case, the relationship between exposure to asbestos and lung cancer is well established. But hold on a minute: 'Did your husband smoke cigarettes?' 'Yes, he enjoyed a quiet smoke, perhaps 20 or 30 a day'.

If my interpretation of the results of Selikoff & Hammond (8) is correct, asbestos exposure alone increases the risk of death from lung cancer 5-fold, but smoking increases it 11-fold, so that the increased risk due to smoking and asbestos exposure was found to be 53-fold. The main cause of the husband's death was his smoking, so the widow should only receive one third of any assessed damages. But why did he smoke? Should not the tobacco companies be sued for advertisements that encouraged him to smoke, and should not his mates be sued because their peer group pressure confirmed him in the habit? Plenty of opportunities here for the 600 000 lawyers Oreffice refers to.

The settlement referred to at the beginning of this chapter was purely a legal and financial manoeuvre, with no scientific backing. In a sense, the scientific and technical community had failed. It had failed to provide clear and unequivocal answers that the public and the lawyers could understand, therefore the settlement is an attempt to resolve the issue outside of the technical area. Rather than face the prospect of bringing the sort of complex technical arguments I have outlined above to court, the chemical companies have tried for a simpler solution.

From this general survey of the topic, and the more detailed analysis of the herbicides problem in Chapter 14, we can draw some general conclusions. Firstly, the scientific community must work hard to get the basic principles of chemical toxicity and exposure over to the public, the media and the legal system. Secondly, it must ensure that a forward-looking plan to increase the safety of chemicals is made, on a broad front. We should not allow an isolated case, just because it is sensational, to drive our planning. There is still a great inertia against doing anything sensible about chemical safety, which is regularly disturbed by a greater hysteria about irrelevancies. Finally, continued effort must be made to develop reliable short-term tests to predict toxicities, to replace the long and wasteful animal experiments we now have to rely on. The theme of this chapter is expanded upon in reference 9.

We have to put science into the legal system we use to control chemicals. We must ensure that good objective science is used to secure justice for the chemically injured and to allow the safe marketing of useful chemicals. I cannot follow society into the Legal World, as it is alien to me, but I can see what our mission must be in the Chemical World.

References

1. Oreffice, P. F. (1984). Law and the threat it poses to the US chemical industry. *Chemistry and Industry*, 2 January 1984, 15–17.
2. Martindale (1982). *Martindale. The Extra Pharmacopoeia*, 28th edn, ed. J. E. F. Reynolds. London: The Pharmaceutical Press.
3. Elkins, H. B. (1982). The real world – or science fiction. *American Industrial Hygiene Association Journal*, **43**, 717–21.
4. Venitt, S. (1980). Bacterial mutation as an indicator of carcinogenicity. *British Medical Bulletin*, **36**, 57–62.
5. IARC (1982). *IARC Monographs on the Evaluation of the Carcinogenic Risk of Chemicals to Humans. Chemicals, Industrial Processes and Industries Associated with Cancer in Humans.* Supplement 4. Lyon: International Agency for Research on Cancer.
6. Epstein, S. S. (1976). Regulatory aspects of occupational carcinogens: contrasts with environmental carcinogens. In *Environmental Pollution and Carcinogenic Risks*, INSERM Symposium Series, vol. 52, ed. C. Rosenfeld & W. Davis, pp. 389–402. Paris: Editions INSERM.

7. Westerholm, P. (1976). Administrative aspects on regulation of carcinogenic hazards in occupational environment. In *Environmental Pollution and Carcinogenic Risks*, INSERM Symposium Series, vol. 52, ed. C. Rosenfeld & W. Davis, pp. 403–16. Paris: Editions INSERM.
8. Selikoff, I. J. & Hammond, E. C. (1979). Asbestos and smoking. *Journal of the American Medical Association*, **242**, 458–9.
9. Hawkins, K. (1984). *Environment and Enforcement: Regulation and the social definition of pollution*. Oxford: Clarendon Press.

16

The natural world: safe or unsafe?

The people of the villages of Karain and Tuscoy in Turkey build their houses from blocks of soft rock hewn from the neighbouring volcanic tuffs. They also use caves in the same rock for storage. This is hardly remarkable; the remarkable characteristic of the villagers is that many die of pleural mesothelioma, which is a very rare disease in most populations. However, in Karain it has accounted for over 40% of all deaths (see the summary account in reference 1). Of course, the only other population we know of that has a high incidence of mesothelioma is that of the asbestos workers (Chapters 8 and 13). Logically enough, examination of the volcanic rock from around Karain and Tuscoy has shown the presence of fine fibres of erionite, a clay-like material. There is therefore a very strong suspicion that the presence of erionite, localised near these two villages, is the cause of the peculiar disease pattern among the villagers. Such a cause and effect relationship is not absolutely proven, but must be accepted as 99% certain. We have therefore an interesting comparison between a circumscribed industrial population exposed to a chemical hazard caused by industrial processes and a small, local population exposed to a similar hazard from a natural source.

The purpose of this chapter is to point out that chemical hazards in the environment are not a novel feature of the Chemical Age. Their nature and relative concentrations have changed enormously. The Chemical Age has removed some hazards and created many others. We therefore have to make many judgements as to benefit and detriment; the good old days were not all good.

From Jean-Jacques Rousseau onwards, we have all cherished a wish to go back to the idealised life of the Noble Savage. Captain Cook and the other explorers of the South Seas thought that they were seeing it. However, some restraints such as food supply, complex social organisation and ritual wars must have regulated the population of the islanders and made life less than ideal. In fact, for most of our life on Earth, existence has been a brutish hand-to-mouth struggle to feed ourselves and just survive in a hostile world (but see Chapter 2 for other views on this). Various forms of social elite managed to raise themselves above the mass and produce a cultural veneer over the brute reality. Today perhaps 25% of the world's population is free from the struggle to find tomorrow's dinner, the rest are, as ever, caught in the battle to survive. Because of this necessary preoccupation with food, we can consider food in the natural world first.

A very interesting account has been given by Gelfand (2) of the food and eating habits of the Shona nation in Zimbabwe. The Shona have been little influenced by western habits, and in rural areas still represent a way of life independent of the Chemical Age (as far as food goes). The bulk of the diet is a cereal: maize, sorghum or millet. This is supplemented by peanuts, other vegetables and fruit when available. Meat of a wide variety is eaten (including mice, caterpillars, locusts) but the principal form is ox or goat flesh which is dried and stored for the winter months. The diet is therefore sufficiently varied and adequate in good seasons but, in bad seasons, the inability to store food for long periods makes the Shona vulnerable. The main objection to the food from the western point of view would be the sheer monotony of the staple, stiff porridge. The Shona eat in silence; they eat to satisfy hunger.

Food storage is of interest, as we will see later. Gelfand says that millet and rice can be stored for years, sorghum for about a year and maize for several months. The latter seems especially sensitive to insect attack.

Most primitive diets consist of one staple, supplemented by whatever else can be obtained. I remember an account of life on the Blasket Islands off the coast of Ireland. The diet was potatoes, fish and milk. In winter, if the gales were bad enough to prevent you launching your boat, then you survived on potatoes. This was an adequate diet in good times. A consequence of the reliance on a staple food is that if the crop fails, you go hungry, or eat whatever you have stored from the last crop. This may already have been consumed by insects, rats or moulds. The residue, which you eat because you are hungry, is contaminated by the chemicals left behind by the previous diners. The residues of rodents and insects are objectionable from a hygienic point of view, since infections may be spread this way. From a chemical viewpoint, the extra protein, uric acid and other chemicals in the insects or their products are not harmful unless you have specific allergies. I know of no chemical toxins resulting from such contamination of food. The situation is different for moulds; they do produce toxic chemicals when growing on foods.

The toxic chemicals from moulds are known as mycotoxins. The best known of the mycotoxins (apart from alcohol) is undoubtedly penicillin, from moulds of the genus *Penicillium*. However, penicillin is much more toxic to bacteria than it is to man, so that it has proved to be a useful tool rather than a bothersome poison. Many other mycotoxins are now being recognised whose properties are solely adverse to human health. The population of the USSR was close to starvation during the Second World War. In order to survive, they used grain of doubtful quality, including that which had not been gathered in the autumn harvest but had lain under the snow all winter. This grain was gathered the following spring and used for making bread. Many thousands, hundreds of thousands of people suffered toxic effects from this bread. In extreme cases the symptoms were: extensive skin damage with damage to the eyes also; internal injury to the digestive tract with bleeding from the mouth; and other symptoms suggestive of interference with the immune system and blood-forming system of the body. Victims died in 4–8 weeks as they wasted away (3). This disease became known as alimentary toxic aleukia (ATA) and was associated with the presence of the mould *Fusarium poae* on the grain, but the chemical nature of the presumed toxin was not known.

Years later in the USA there were incidents of cattle showing toxic symptoms after feeding on mouldy corn (maize). The practice in the midwest is to bring cattle off the range and pen them in large numbers to fatten them for slaughter. They are fed with corn from large bulk stores, which is delivered into feeding troughs or runnels alongside the pen by mechanical shovels or dozers. The cattle can put their heads through the wire fence to reach the troughs and feed. Because the grain is handled in large quantities it is not surprising that areas of fungal contamination are not noticed. Subtle effects of contamination may be ‘failure to thrive’, obvious effects are diseases associated with bleeding, similar to ATA. The mould *Fusarium tricinctum* was isolated from fungal contaminants of grain implicated in one episode of poisoning in Wisconsin (4). From this was isolated a toxic chemical, T-2 toxin. The implication is that T-2 toxin is the cause of the cattle disease and, further, that it was the cause of ATA in the USSR. Actually, T-2 toxin is one of a family of toxins, the trichothecene class of mycotoxins, therefore the various stock or human diseases could be caused by other, related, trichothecenes or by a combination of several of them.

The back-to-nature purist will now say: ‘Your example simply shows the disadvantages of modern, intensive agricultural practices; it would not happen if the cattle were raised naturally on the range’. True enough – however, the incidence of human disease related to the prevalence of specific fungal infections of corn grown under traditional practices in southern Africa is described by Marasas (5). Similarly, the relationship between human and animal diseases and the occurrence of mould on food stuffs is described for a number of advanced and primitive communities (6).

Mycotoxins that cause cancer are discussed in Chapter 8. Since the discovery of aflatoxins and the isolation of T-2 toxin, much work has been done on mycotoxins in stored products. There is no doubt that they have an adverse effect on health but these effects are often subtle and not obvious until a dramatic episode of poisoning occurs. Mycotoxins are part of the natural world, but do present a chemical threat to humans.

We are more familiar with the large fungi such as mushrooms and toadstools than with the moulds which insidiously contaminate our food. Mushrooms of various kinds are eaten quite commonly by rural communities and why not? We are told that the majority of mushrooms and toadstools are edible. It is the minority that causes nasty surprises, even among communities that commonly eat mushrooms and are experienced in distinguishing edible from poisonous ones. In 1963 I was touring in northern Italy and not paying much attention to news except for what I read on the backs of other people's newspapers in cafes. For a number of days there was a series of headlines reading '*X morte alle funghi*', in which the numeral *X* increased each day, to about 30 if I remember correctly. Such mushroom poisonings once again illustrate that nature has its own arsenal of chemical weapons with which to slaughter the unwary.

Other toxic chemicals are produced by plants themselves, in fact their number is legion, and they are best considered as drugs (Chapter 9) since many have found uses (or abuses) in that role. Besides being directly toxic, plant chemicals can have more subtle effects such as teratogenicity, e.g. they may cause defects in the foetus when eaten by the pregnant mother. For example, the alkaloid anagyrine from lupins causes 'crooked calf disease' in cattle if eaten by the dam. There is no certain case of plant teratogens acting on humans, but an interesting possibility is discussed by Kilgore (7). A baby boy was born with deformities to his arms and hands in a rural area of California. Goats had given birth to kids with deformities at the child's home, as had a bitch. The mother was worried by the possibility that spraying of 2,4-D had caused the deformities, but spraying had only occurred some miles away more than a year before the child's conception. The kids and puppies had deformities similar to those of crooked-calf disease. The browsing areas of the goats were examined and found to contain a perennial lupin as a substantial part of the forage. This lupin was found to contain anagyrine; when seeds from the lupins were fed to a lactating goat, anagyrine and other alkaloids appeared in the milk. The child's mother and the bitch had both drunk the goats' milk during pregnancy. This circumstantial evidence suggests strongly the causal chain described below.

> Goats eat lupins containing anagyrine, the alkaloid passes into the goats' milk, mother drinks goats' milk and exposes herself to anagyrine, anagyrine in the maternal circulation causes defects in the developing foetus, a deformed child is born.

This cannot be regarded as rigorously proved, yet the evidence is strong. Compare with the apparent non-involvement of 2,4-D, often convicted on much weaker evidence (Chapter 13). Other aspects are worrying. The goat and its milk is a fashionable animal in the back-to-nature cult, but it is a wide forager apt, as above, to pick up strange contaminants, and is a voracious destroyer of the environment. I have seen a feral goat attempt and fail to reach the leaves on a eucalyptus sapling. The goat's answer to this problem was to back up to the sapling, catch it in the crook of one horn, bend its head back until the sapling cracked, then munch the fallen greenery. Ingenious but destructive.

Although leguminous plants (peas, beans etc.) are a source of much human food, some varieties also contain quite potent poisons. In addition to some strains of lupins (as in the account above), the peas of the genus *Lathyrus* produce a toxin which attacks the nervous system and causes paralysis of the lower limbs, the disease 'lathyrism'. This disease is endemic in parts of India (8) in which the seeds are eaten as 'khesari dal'; they are eaten because, although the cause of the disease is well known, the plant is a very successful crop and thus forms an inevitable part of the diet of those whose poverty gives them no choice. Some strains of ornamental sweet pea also contain a toxin, this time one which interferes with bone growth. I am careful now not to grow sweet peas alongside the snow peas we grow for the kitchen. My children are fond of picking and eating the snow peas, and they might not distinguish the hairy pods of the sweet pea from the smooth, innocent ones.

These brief examples illustrate the point that the natural world is full of chemicals which affect humans through their food, medicines and directly from the environment. This has always been so, and the burden of resulting disease has been accepted as normal, a contribution to the background of general ill health and minor complaints that occur in any community, and to the 2% or so of deformities that occur among children at birth (Chapter 7). A close analogy can be made with radiation. Low-level radioactivity exists in all natural things, mineral, vegetable, animal. People have always been exposed to radiation and have eaten radioactive materials. This again has probably contributed to minor disease in the 'natural' state. Obviously it is sensible to limit the exposure of people to radiation since an increase in dose means an increase in effect. However, radiation is not new; there is a difference in degree between natural and artificially induced radiation, not in kind. Similarly, toxic chemicals are not new, the synthetic ones are increasing greatly in use and variety, but the synthetic chemical does not differ from the same chemical from natural sources.

Many of these points are covered elsewhere in this book, but I have put in this chapter towards the end to redress the imbalance that may have occurred as a result of a lot of discussion of synthetic chemicals. We, in the natural world, are not entirely safe from chemicals.

References

1. Wagner, J. C., Berry, G. & Pooley, F. D. (1980). Carcinogenesis and mineral fibres. *British Medical Bulletin*, **36**, 53–6.
2. Gelfand, M. (1981). Cancer among the Shona in relation to their traditional diet and medicines. In *Dietary Influences on Cancer: Traditional and Modern*, 1st edn, ed. R. Schoental & T. A. Connors, pp. 11–35. Boca Raton, Florida: CRC Press.
3. Joffe, A. Z. (1971). Alimentary toxic aleukia. In *Microbial Toxins, Vol. VII. Algal and Fungal Toxins.* ed. S. Kadis, A. Ciegler & S. J. Ajl, pp. 139–89. New York: Academic Press.
4. Bamburg, J. R., Riggs, N. V. & Strong, F. M. (1968). The structure of toxins from two strains of *Fusarium tricinctum*. *Tetrahedron*, **24**, 3329–36.
5. Marasas, W. F. O., Wehner, F. C., van Rensburg, S. & van Schalkwyk, D. J. (1981). Mycoflora of corn produced in human esophageal cancer areas in Transkei, southern Africa. *Phytopathology*, **71**, 792–6.
6. Ueno, Y. (ed., 1983). Toxicoses, natural occurrence and control. In *Trichothecenes. Chemical, Biological and Toxicological Aspects*, 1st edn, pp. 195–307. Amsterdam: Elsevier.
7. Kilgore, W. W., Crosby, D. G., Craigmill, A. C. & Poppen, N. K. (1981). Toxic plants as possible human teratogens. *California Agriculture*, November 1981.
8. Rutter, J. & Percy, S. (1984). The pulse that maims. *New Scientist*, 23 August 1984, pp. 22–3.

17

Religion, food and chemicals

No, I have not finally gone nuts. I can defend this association of oddly assorted topics, and will do so if you have the patience to follow through a fairly long argument.

Religion is a basic element of human societies. This is purely a scientific observation, and I will not argue for or against religion, but simply point out that all societies have had a religious element. Many years ago I ploughed through the abridged version of Sir James Frazer's *The Golden Bough* (1) and became sufficiently convinced of the universality of religion not to need to read the full twelve volumes. Since Sir James's time the study of comparative religion has improved, but the one point I wish to emphasise is that all societies have religion and therefore it must fill a basic human requirement. There are two aspects to religion which fill differing needs. First, there is the comforting aspect of belief in a supernatural protector, which can give meaning to life, comfort in death and identification of a place for people in the natural and supernatural world. From this follows a sense of reassurance; if you do the right thing the gods are happy and you will prosper. Conversely, of course, if you do not prosper, then you have offended and must repent. The second aspect of religion is the ceremonial one. People have a yearning for ceremony, which certainly can be filled by secular activities, but is very often satisfied by religious ceremony. The great religious reforms have attempted to eradicate ceremony, but it creeps back. Thus the Protestant churches have removed the more flamboyant practices of Roman Catholicism, but the ceremony is still there, if conducted in dark suits rather than crimson vestments. Very few of the ceremonies in Christian churches have any justification in basic Christian dogma, but cut them out and you lose half your congregation.

Formal religion has declined in observance over the last 100 years or so in western countries. This decline can be ascribed partly to the upsets in society caused by the Industrial Revolution and partly to the growth of scientific knowledge, which offers an alternative explanation of phenomena formerly thought to be under supernatural control. The changes in society are fairly obvious; the European village was centred around the inn and the church, the axes of the secular and the spiritual community. Somewhere nearby was the manor house, chateau or schloss, the home of the temporal power. This tight community has been broken up, so that the ceremony and reassurance of a small religious unit is not available to very many socially mobile city dwellers. At the same time as social forces were breaking up small communities, the growth of science was offering an argument to justify the diminution in the status of religion that was occurring. Other non-religious philosophies were also displacing religion; however, one may wonder whether or not Marxism is another opiate, and hence a religion by another name. So strong is the religious need that the displacing philosophies may themselves take on many of the characteristics of that which they supplant. Whatever the mechanism, I believe that the decline in organised religious observance has left a psychological void in modern people. Something has to fill this void, and if the rational is too cold and unemotional, then it has to be the irrational. People may attempt to fill this need through orthodox and rational channels, which is why the doctor becomes the dispenser of Valium (Chapter 9) in place of the priestly wine and bread. Others may turn to various irrational cults, ranging from odd diets through herbal medicines to downright old-fashioned Satanism.

Now we come to food. I have chosen the various irrational attitudes to food to illustrate what I see as the religion-substitution syndrome. It would be just as easy to use drugs as the example (which I have done to a degree in Chapter 9), but I have chosen food because this topic has not featured elsewhere in this book (except for minor aspects discussed in Chapters 8 and 16). Lastly, we can bring in chemicals, as they are the villains in the popular conception of modern food.

If we look at the various diets that are proffered to us by the media in great profusion, we find they fall into two classes; those for healthy people to keep them healthy, and those for persons suffering from a disease (which may be real, or imaginary). Of the latter class, those for victims of cancer are the most numerous. The imaginary diseases of the second class really blend into the first class, for healthy people often need the excuse of a supposed health problem to justify experiments in diet. Let me also say that I can see rational elements in many or all of these diets, which are mixed with greater or lesser proportions of the irrational.

When we examine the foods that are generally eaten in developed countries we find there are three sources of possibly toxic chemicals; those that are naturally present in food, those that have accidentally contaminated it, and those additives which have been added after harvest to preserve or improve the food in some way. The naturally occurring toxic chemicals in food have been mentioned in Chapters 8 and 16 and I will not discuss them further. They are summarised in Table 17.1, which is taken from a book by Farrer (2).

The chemical contaminants enter food from the environment during the life of the plant or animal (pollutants, residues of pesticides, etc.), or are absorbed from packaging material during transport and marketing (e.g. vinyl chloride monomer from PVC containers) or may actually be formed within the food during processing or cooking (see the comments on PAH and nitroso compounds in Chapter 8). Where the contamination has been gross, the effects can be dramatic, as witnessed in the Spanish cooking oil episode (1981) in which 300 persons died of eating contaminated oil; there are still doubts about the cause of the poisonings (3) but the result was appalling. Similar tragedies have happened as a result of the gross contamination of grain by concentrates of organophosphorus pesticides, and from other accidents in which toxic chemicals have carelessly been allowed to mix with foodstuffs. The remedy for these incidents is obvious, and they are probably not the type of contamination which most concerns people. It is the possibility of widespread contamination by a chemical which slowly causes toxic effects which is most feared. The spread of organochlorine compounds throughout all food chains, as a result of the use of DDT, benzene hexachloride (BHC), dieldrin and so forth as insecticides, was the event that raised public concern. The chemicals are present in human fat and must have reached us mainly from our food. The contamination of human fat by particular compounds has declined (4) as these pesticides have been withdrawn from agricultural use.

The possibility of toxic effects arising from low-level contamination of food by a very wide range of chemicals has been discussed. I can take just two examples at random from many. The first (5) is an evaluation of the risk incurred by allowing dishes to drain and dry, rather than rinsing and wiping them. Detergent liquids are slightly toxic (the oral LD_{50} of various detergents is estimated to be between 1000–3000 mg/kg) but the probable human intake is very low. The article (5) concludes that it would be sensible, on gastronomic rather than toxicological grounds, to rinse or wipe dry the dishes before putting them away. My second example (6) discusses the risk arising from the contamination of food by anticoagulant chemicals, which may be present in certain foodstuffs as a result of their use as pesticides. They are mainly used to control rodents but, strangely enough, cattle in Latin America are treated with these chemicals to control vampire bats. Dracula beware! The authors do not come to a firm conclusion as to

Table 17.1. *A classification of naturally occurring toxicants in foods*

Type		Examples	Made safe by
I	Toxicants concentrated in foods by vectors	Andromedotoxins and other substances in honey	Abstention
		Tremetol in milk	Abstention
II	Toxicants concentrated in organisms used for food	Ciguatera poisoning in fish	Abstention
		Saxitoxin in shellfish	Abstention
III	Toxicants occurring normally in plants and animals used for food		
	(i) At all times		
	(*a*) Rapid action	Cyanide in bitter cassava	Leaching and fermentation; selective breeding
		Cyanide in lima beans	Limited use in a mixed diet; selective breeding
		Cyanide in apricot kernels	Very limited use
		Alkaloids in bitter lupins	Leaching; selective breeding
		Vicine in broad beans	Use in a mixed diet
		Oxalyl-substituted amino acid in chick peas	Use in a mixed diet
		Oxalic acid in rhubarb and spinach	Use in a mixed diet
		Nitrates in spinach	Used in a mixed diet
	(*b*) Slow acting	Goitrogens in brassicas	Iodine supplementation
		Carcinogens in bracken	Abstention
	(ii) Only at certain times	Puffer fish poison in fugu	Great care in preparation
		Hypoglycin in akee	Eat ripe fruit only
		Solanine in potatoes	Avoid peel, eyes, shoots and 'greened' tubers
IV	Toxicants formed by interaction of food constituents with		
	(i) Other food constituents	Nitrate in vegetables via nitrite to form nitrosamines with secondary amines	Use in a mixed diet
		Nitrite from certain preserved foods with secondary amines to form nitrosamines	Use in moderation and in mixed diets

Table 17.1. (*contd*) *A classification of naturally occurring toxicants in foods*

Type		Examples	Made safe by
		Anti-vitamins with vitamins to destroy the specific protective effects	Eating a normal varied diet
	(ii) Medically prescribed drugs	Active amines in cheese, chocolate, etc. with monoamine oxidase inhibitors	Abstention under medical advice
V	Allergens	Idiosyncratic reaction of individuals to certain foods	Abstention based on experience and medical advice

See (2), with permission of Dr Farrer and Melbourne University Press.

the risk involved, due to lack of data on dosages, but advise caution particularly for pregnant women. They suggest that physicians should be aware of this possible problem.

There are very many chemicals that contaminate food, and we will find more as we improve our analytical methods further (Chapter 6). There is no evidence that the majority of these chemicals, at the concentration levels found in food, are at all harmful to us. I admit the difficulties of assessing very long-term declines in health and assigning them to a cause, therefore we should reduce food contamination as much as we can, learning from those mistakes that have in the past allowed chemicals to permeate through the environment to our food. Accidents are always possible; it is only through educating the public that we can avoid gross contamination of food. Toxic chemicals have to be recognised as such and kept away from food.

Now we come to the food additives. These are put in food for a number of reasons, and include colours, flavours, sweeteners, preservatives, modifying and conditioning agents, and others. It seems to me that it is possible to separate out those additives which have a justifiable function from those which are purely cosmetic. Preservatives have a function, particularly in a complex, specialised society, and their use is justified, provided that there is no risk in their use (see Chapter 8). Some preservatives border on the cosmetic; the familiar appearance of bacon and ham largely depends on nitrates. Conditioning agents have a use if the public wants a particular product; for example if you want to buy frozen pastry that you can store in your freezer against the time your spouse demands apple pie, then you can have it, but the manufacturer has to modify the normal mix to get a product that will be workable after freezing and thawing. Colours, flavours and sweeteners are less easy to justify. The taste for sweetness is an acquired one, and can be lost again if one wishes so to do. Sweetness and the other cosmetic factors are part of the competition between manufacturers to woo the shopper to their product. Sensible regulation which controls additives will keep competitors on an equal footing in this regard, and reduce the temptation to introduce more and more additives. Two views on additives are given by Farrer (2) and Millstone (7). The former presents the more conservative view, whilst Millstone is concerned that the use of additives is growing fast whilst we still know little about them. He therefore presses for more research to acquire information, and more effort to make that information available to the public. I feel the manufacturer has to justify the additives used, and it must be in terms other than that the customer demands the product. I also feel a little sad about the customer who will elect to buy potato crisps or corn chips flavoured with an exotic and obviously artificial flavour. What is the next taste treat? Papuan long pig? Cantonese chow?

There is some recent concern about the residues of antibiotics and sex hormones in meat and poultry. The chemicals are given to the living animals to increase body weight and thus the yield that the producer can market. It seems to me that this practice should be watched carefully, for such usage may prejudice the usefulness of antibiotics in their prime value to society as controllers of disease, and because sex hormones are part of a control system within our bodies which may easily be upset. Let us use chemicals in the areas of most social benefit.

Perhaps the best way to assess our foods is to look around us. The present generation of children and young adults is as healthy as any previous one, if not more so. The only apparent dietary problem in the richer countries is excess. In the other two thirds of the world it is scarcity.

Let us go back to the second of the two classes of diet we left some time ago, and discuss the various diets proposed to either ward off, or help cure, cancer. Let me emphasise first that the basic three rules of nutrition apply just as much in sickness as in health. The three rules are; moderation, variety and balance. The average person can achieve this without reference to detailed instructions, and the traditional diets of most cultures follow these rules as far as they can (there may have to be excessive reliance on one staple food). A variety of cereal grains, fruits, vegetables, animal flesh and dairy products eaten in moderation is adequate. A diet selected to reduce the likelihood of cancer must be based on the same ingredients, with attention to the reduction of those factors known or suspected to increase the chances of cancer formation. This involves reducing those factors we have discussed here and in Chapter 8, namely salt-cured and smoked meats, animal fats, those foods naturally contaminated with carcinogens (e.g. aflatoxins), and excessive alcohol. It should be noted that supplementation of normal diets by added chemicals or 'tonic' foods is not normally necessary. If the person is in a diseased state, then such additions to diet do become necessary. Also note that such diseased states are defined and diagnosable. The purveyors of extra vitamins, supplements and tonics first convince you that you have a disease you were not aware of, then sell you the remedy. We are not always on top of the world, but chemicals will not elevate us to that lost eminence.

The situation is not greatly different for the person actually suffering from cancer. It is necessary to get the patient to eat an adequate and varied diet, which may need to be adjusted for specific needs and for disabilities caused by the disease. To counter the wasting effects of cancer, it is often necessary to provide additional protein in digestible form. Nutrition is an important part of orthodox cancer treatment and if it is being neglected, then it ought to be cultivated. However, there is a school of alternative nutritional practices for cancer patients which is a mixture of quite illogical practices and downright quackery.

Herbert has reviewed nutrition cults in general (8) and alternative cancer nutrition (9), in particular, pointing out the lack of sound support for the notions advanced, and the common features of the proposed diets. These include the exclusion of flesh and dairy products, thus robbing the cancer victim of absorbable iron, calcium and the most useful source of protein. In substitution the patient receives a large quantity of vegetables, fruit and cereals, which do not supply the missing requirements, and may further compound the deficiencies. Thus a diet high in cereal and fibre may restrict the uptake from the gut of what little mineral content is in the diet. The diet is also supplemented by a number of chemicals, including megadoses of vitamin C, supplements of vitamins A and E, selenium (a trace element of which 50–200 μg daily are safe and adequate), various ill-defined and unrecognised B vitamins (e.g. vitamin B_{17} or laetrile, vitamin B_{15} or pangamic acid) and other substances. None of these have been shown to have value in treating cancer patients, when the tests have been carefully performed on a comparative basis. Some patients taking such substances in their diet will have recovered from cancer; others will have got worse. There is no evidence that these nutritional practices have been the reason for those improvements which have occurred. One also hopes that they have not contributed too markedly to the decline of health in the other patients.

The practices that Herbert has argued to be faddism and quackery are well displayed in the book by Kidman (10), which extols their virtues. All the dieting restrictions and additions mentioned in the previous paragraph are recommended by Kidman, with no comprehensible explanation or meaningful arguments. Thus we find (p. 107 of ref. 10) that beansprouts cleanse the bloodstream and create alkaline conditions to enhance the action of the pancreatic enzymes which play an important part in digesting the mucous coat of the cancer cells. We are not told how the beansprouts cleanse the bloodstream, nor even what this phrase actually means. The acidity/alkalinity of blood and other body fluids is well controlled, and unlikely to be altered by any but massive intakes of sprouts. The pancreatic enzymes are a favourite of Kidman's, but a bit of bother too, for she totally fails to realise that they are given out into the gut, which is their site of action, and cannot get back into the body. Therefore they cannot contact any cancer except those actually on the inner wall of the intestine. They certainly are not circulating in the blood. This is also why taking pancreatic enzymes by mouth (also advocated by Kidman) is useless, particularly as they themselves are liable to digestion by enzymes normally present. Giving such enzymes by enema is also useless, for exactly the same reason that they cannot enter the body from the intestine as complete proteins. If enzymes given by the mouth or by the anus are digested (unlikely in the latter case) the products are much smaller molecules which do not have the activity and properties of enzymes. Kidman advocates a variety of enzymes, vitamin excess, selenium, laetrile, etc., in fact a regular vocabulary of nostrums now being enshrined as the orthodoxy of the alternative.

We can condemn these dietary practices as unscientific and irrational. To an extent we can excuse their advocates for lack of knowledge. What is disturbing, however, is the complete lack of any logic in the selection of the diets. The faults in Kidman's book (10) are not merely in scientific knowledge, they are self-evidently illogical. It does not require scientific knowledge to find the internal contradictions in the book; any intelligent layman can do it. Thus we are told (p. 127) that coffee, tea and chocolate should be given up because they are stimulants which also burden the liver. Fair enough, there is truth in this. But then the sufferer is told to take coffee enemas whenever he/she feels the need (p. 147). Coffee is taken to the liver, where it stimulates the production of bile, which then carries waste substances out of the body. The reader is not told why coffee arriving at the liver by a normal route is bad, but is good when it arrives by a rather perverse route. Nor is the reader told that coffee enemas may be fatal (11).

Kidman's book is only one expression of the large and complex world of nutrition cults. I have no doubt she is sincere in her beliefs, and here is the intriguing aspect of this phenomenon. The many persons who make money out of selling or manufacturing quack foods and supplements are understandable, even if inexcusable. They exercise the basic right: 'Everyone has the right to make a killing'. But many others sincerely believe in illogical rubbish. Why? Because there is some basic trend in society towards obscurantism. Why? I am not totally sure, but feel it has something to do with the innate religious impulse we discussed at first.

You are probably about to accuse me of wandering a long way away from chemicals. Not so, for the interesting conclusion from the above considerations is that an alternative cult is advocating the use of chemicals in food, in excess of any orthodox recommendations. Laetrile, which can generate cyanide and probably has poisoned a number of people, is a favourite chemical. So is selenium, which appears to have some value in protecting against cancer, but at near-toxic doses – not a chemical with which you should freely treat yourself. Similar considerations apply to vitamin A.

We should question the additives put in our food by the manufacturer purely for cosmetic reasons. Equally, we should question the chemicals added to diet by the persistent advocates of dietary cults. In each case we must use reason to arrive at a conclusion. In the first case reason must contend commercial pressures. In the latter, it must contend against non-reason.

The parallelism between the discussion on food and that on drugs in Chapter 9 is quite obvious. We must conclude that large sections of the public have no objection to chemicals as such; rather they prefer them in complex and uncharacterised packages rather than in defined and tested single lots. It is the cultural ambience of the chemicals which is significant, not the fact of being chemicals.

The two connections between faddism in foods and the religious impulse are thus the common rejection of reason, and the shared adoption of ceremony. Ceremony is a part of normal cooking and eating rituals, as evidenced by the formal dinner party; it has an added role in faddism, where one suspects that the ritual is vital to the expected benefit. Grinding one's own corn, culturing your own yogurt or brewing herb teas are activities which have a ceremonial side beneficial to the performer and hence are related to the need for religious ritual. I do not say that these activities are without benefit, but I want you to distinguish between the satisfaction of psychological needs, and the possible nutritional advantages of these activities. The latter are very largely imaginary; the former is related to the religious makeup of people. I also add that I am not against religion, but am warning against mixing rational science and religion.

There is no rational way of proving the existence of God; by contrast one should not use faith as the justification for the nutritional value of food. Faith in religion and logic in science can coexist, each in their own sphere. It is only when we neglect the boundaries that we have problems.

A rational approach to food and the chemicals which constitute it is the only way that we can control abuses of our food by the producer intent on profit or the retailer who wishes to deceive. A retreat to an obscure world of private faddism will help no one. The production and supply of food must of necessity become yet more complex as society becomes more urbanised. We cannot reverse this trend, but we can maintain the quality of food if we face the problem squarely.

Nutrition is a science, not a religion. It is the study of the utilisation of a group of chemicals functionally known as foodstuffs. Whether obtained from the farm or factory, all foodstuffs are chemicals and are amenable to conceptual treatment as chemicals.

References

1. Frazer, Sir James (1922). *The Golden Bough.* Abridged edition. London: Macmillan.
2. Farrer, K. T. H. (1983). *Fancy Eating That.* Melbourne: University Press.
3. MacKenzie, D. (1984). Spanish poisoning: was cooking oil really to blame? *New Scientist*, 18 October 1984, 10.
4. Abbott, D. C., Collins, G. B., Faulding, R. & Hoodless, R. A. (1981). Organochlorine pesticide residues in human fat in the United Kingdom 1976–7. *British Medical Journal*, **283**, 1425–8.
5. Anonymous (1984). Do detergent residues damage the gut? *The Lancet*, 18 August 1984, 384.
6. Engeman, R. M. & Pank, L. F. (1984). Potential secondary toxicity from anticoagulant pesticides contaminating human food sources. *New England Journal of Medicine*, **311**, 257–8.
7. Millstone, E. (1984). Food additives: a technology out of control? *New Scientist*, 18 October 1984, 20–4.
8. Herbert, V. (1980). *Nutrition Cultism, Facts and Fictions.* Philadelphia: George F. Stukley Company.
9. Herbert, V. (1984). Faddism and quackery in cancer nutrition. *Nutrition and Cancer*, **6**, 196–206.
10. Kidman, B. (1983). *A Gentle Way with Cancer.* London: Century Publishing.
11. Eisele, J. W. & Reay, D. T. (1980). Deaths related to coffee enemas. *Journal of the American Medical Association*, **244**, 1608–9.

18

Finding out about chemicals

I have emphasised repeatedly the need to base any comments or decisions about chemicals upon facts, not upon prejudice, fancy or propaganda. You as readers are therefore quite right to ask how you are to secure the facts. Therein lies one of the key problems in the whole business. I admit that the facts are not freely obtainable by the layperson, and that this is largely the fault of the scientific community. However, if the layperson is given some guidance, he or she can hunt out a lot of information. This chapter is intended as a brief guide to chemical information. A valuable help would be a cooperative librarian and, in my experience, librarians are very pleased to help in answering intelligent questions. At least our library staff have managed to maintain a cheerful appearance during my very frequent visits.

In Fig. 18.1 is a representation of the interface between you as the inquirer and the primary sources of information. These latter may be articles in scientific journals describing the synthesis of a chemical or the estimate of its toxicity, or they may be reports from industrial laboratories or applications for patents or many other types of information, some of which are listed in Fig. 18.1. This primary information is very specialised and largely incomprehensible to the reader who is unfamiliar with the jargon and the concepts which are involved. An article in a scientific journal will have an introductory paragraph designed to show how the work to be described fits into the more general chemical background. This introduction is written for chemists, however, not for laypersons. Government reports may have more extensive background material in them, depending on the expected readership. In general the primary sources of information require interpretation. Also, it must be remembered that each item is very specialised and, by itself, may not have much meaning. If often takes a number of studies by different people on the same topic to produce a synthesis of varying ideas which together have a real value. One research team will write up its synthesis of compound *X* as the only possible way to make the chemical. Another team will have different ideas. Only after reading several articles on the synthesis of X will the newcomer be able to form an opinion as to the best synthesis for him to use. Synthesis is not too controversial; when you select toxicity as a topic, then you have to be very much more careful in interpreting primary information.

If primary information is difficult to use, what do you do? You have the choice of five forms of secondary information, which vary in the degree to which the primary information has been digested or reviewed. Firstly, there are 'reviews'; these range from those published for scientists which examine one field intensively in technical language to those which are deliberately written for the layperson. This type of information can be subject to the bias of the reviewer, and must be assessed with that possibility in mind. The next form of secondary information (B of Fig. 18.1) is that which has been collected from primary sources and collated or sorted, but is presented as factual information, with no attempt at assessing

Fig. 18.1 Pathways by which you as the informed layperson may seek chemical information.

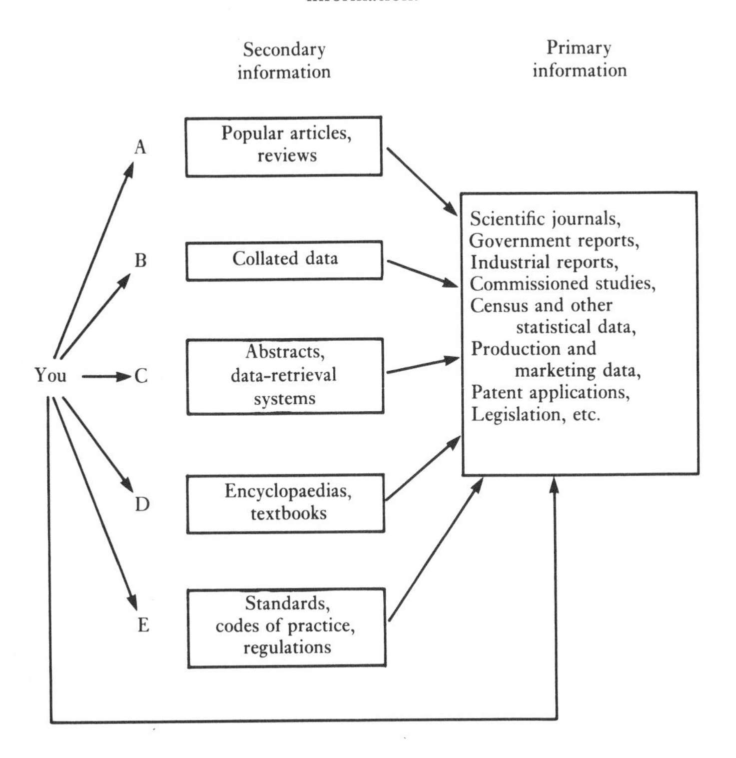

or reviewing it. This is subject to little bias; some is still possible, e.g. data can be deliberately left out. Thirdly, we can have secondary information which is simply a compressed form of the primary data, gathered by a retrieval system. This is merely an expression of the original, and once again bias in presentation is unlikely.

I have put textbooks in a subcategory (D) as sources of secondary information for two reasons. Firstly, the information in them can be a little out of date as it takes more time to publish a textbook than a short review. This may not matter if you are seeking well-established information in an area which is not growing fast, but may be important if you are attempting to follow a topical controversy. Secondly, textbooks often do not go back to primary information, but cite other sources of secondary information. Textbooks are subject to any bias of the author, and may be written in any style from clear to impenetrable.

The last category of secondary information (E of Fig. 18.1) includes all the material relating to legislative requirements and regulations. It is therefore a specialised area which you would consult only if you were directly interested in regulations and controls. The basic technical information on which such regulations are built can be found in other forms of secondary information. Every country has an act or acts of the legislative body which sets the form in which regulations are presented, for example the Toxic Substances Control Act, 1976, of the United States. Such legislation affects the transport of chemicals, industrial hygiene practices, permissible chemical residue levels in food, permissible levels in the environment, and many more related matters.

Matters will be clearer if I give examples of the forms of secondary information. Among reviews written for the scientist are those listed in Table 18.1. I emphasise that this is a random selection, there are many more. These reviews are not really useful to the non-specialist. Of much more use are the articles and reviews written specifically for a wide audience. These may be included in a periodical or journal which also has technical articles, or in a totally 'popular' journal. Examples of the former are *Nature*, which includes newsy and review articles along with scientific articles on molecular biology or cosmic chemistry, and the periodicals of learned societies, such as *Chemistry & Engineering News* (American Chemical Society), *Chemistry in Britain* (The Royal Society of Chemistry), *Chemistry and Industry* (Society of Chemical Industry) and *Chemistry in Australia* (The Royal Australian Chemical Institute). The periodicals *New Scientist* and *Scientific American* are examples of magazines designed to bring scientific knowledge and news to the non-specialist. There are a variety of what may be termed 'science glossies' appearing on the US scene, such as *Omni* and *Omega*, which pay much attention to the pictorial presentation of science to the layperson. In a slightly different class again is *The Sciences*, a publication of the New York Academy of Sciences.

Examples of the sources of collated chemical data are given in Table 18.2 The *CRC Handbook* (1) and the *Merck Index* (2) list the main chemical and physical properties of compounds, and are good starting places for information on a particular chemical. The *Polymer Handbook* (3) is a more specialised source of data on the properties of polymeric compounds. The next two contain compressed information on the toxicity of chemicals. RTECS (4) is updated quarterly and is probably the most extensive source of information on toxicity. Martindale (6) gives much information on drugs; the entries are extensive and it borders on being encyclopaedic rather than a collation.

The prime example of compressed or abstracted data is *Chemical Abstracts* (CA), which abstracts and indexes information from chemical journals originating from all over the world. The resulting volume of data,

Table 18.1. *A representative list of review series in chemistry and related topics (this is a small sample from many available)*

Series title	Publisher
Annual Reports on the Progress of Chemistry	The Chemical Society, London
Annual Review of Physical Chemistry (also includes series on materials science, biochemistry, pharmacology, toxicology and other topics)	Annual Reviews Inc., Palo Alto
Advances in Heterocyclic Chemistry	Academic Press, New York
Progress in Physical Organic Chemistry	Wiley-Interscience, New York
Pharmacological Reviews	Williams & Wilkins, Baltimore

Table 18.2. *Examples of sources of collated chemical data*

Title	Reference
CRC Handbook of Chemistry and Physics	1
The Merck Index	2
Polymer Handbook	3
Registry of Toxic Effects of Chemical Substances (RTECS)	4
Dangerous Properties of Industrial Materials	5
Martindale. The Extra Pharmacopoeia	6

even in abstract form, is enormous. The issues of CA for 1983 (volumes 98 and 99) take up 2.2 m of shelf space in hard copy form. Handling and storage in this form is becoming too much of a burden, so that computer-readable forms of CA will largely displace the hard copy.

The whole business of finding out will become easier to follow if we take a real example and follow it through. On the 3 December 1984 about 30 tonnes of methyl isocyanate (MIC) leaked from the Union Carbide chemical plant in Bhopal, India, and killed some 2500 nearby residents. Suddenly we want to know about methyl isocyanate.

In the earliest press reports, the name was spelled as 'methyl isocynate'. Fortunately, this is a nonsense name, and it was fairly easy to guess that 'isocyanate' was meant. The lesson for the layperson must be remembered, however. Check the nomenclature very carefully before you start hunting for information.

My only previous contact with isocyanates was with toluene diisocyanate (TDI) which is used in the manufacture of flexible polyurethane foams. The vapour of TDI at high concentration is lethal; at lower concentrations (such as might occur in a workplace) it can provoke a hypersensitive, asthma-like attack in 2–5% of normal persons. This sensitisation has a permanent effect and apparently can cause further asthma attacks, even after exposure to initiators other than TDI.

We start our information search with an encyclopaedia as one source of secondary information (Fig. 18.1), namely the index of the *Kirk–Othmer Encyclopedia of Chemical Technology* (7). There are several text references shown in the index. The first tells us the method of preparation of MIC. The next three references are to a general section on organic isocyanates, in which MIC does not figure greatly. However this section gives the physical properties of MIC, including the boiling point of 38 °C and flash point (open cup) of −7°C. It is therefore a flammable and volatile liquid. This section also contains the important information that MIC is produced by Union Carbide and used for the production of several insecticides and herbicides, including 1-naphthylcarbamate, trivial name carbaryl, tradename Sevin. A production of 23 000 tonnes of MIC was predicted for 1980. The health and safety aspects of isocyanates are also given in this section, but mainly relate to TDI and similar compounds. No direct information is given on MIC. The last two index references relate to the use of MIC in the synthesis of a pharmaceutical drug (cimetidine) and a herbicide (tebuthiuron).

This is a start in our search, but obviously we need more information, particularly on toxicity. Let us go directly to RTECS. One's first reaction is to recoil in angry incomprehension at the degree to which the data are compressed and codified. However, they can be decoded fairly easily. The first lines are identification data, including various synonyms and the CAS (Chemical Abstracts Service) Registry Number. The latter is a number accorded on a systematic basis by the CA Service (American Chemical Society) to give an unambiguous identification of a chemical. Then follow the toxicity data, in the order of: route of administration; test animal; mode of expression of toxicity; numerical result with units; and, on the opposite side of the page, a reference to the primary source of the data. This has reduced the title of the scientific journal to a series of letters in a system named CODEN, with the reference details in order of volume, page, year. Thus the third entry:

'ihl-rat LC 50:5 ppm/4H ATXKA8 20,235,64'

expands to

> Inhalation route, rat, lethal concentration causing 50% deaths is 5 parts per million breathed for 4 hours. This information is from *Archiv für Toxikologie*, volume 20, page 235, of 1964.

The bottom of the entry has data concerning standards and regulations.

The relevant information about MIC that we can cull from the entries is that the acute oral LD_{50} is 69 mg/kg for rat and 120 mg/kg for mouse. When inhaled, the LC_{50} for rat is 5 ppm breathed for 4 h (equivalent to 2800 mg/m^3/min, making some assumptions). The LCL_0 (lowest published lethal concentration) is 37 mg/m^3/h for mouse or guinea pig. An oral toxicity of around 100 mg/kg, if applicable to man, puts MIC in the class of moderately toxic chemical (Table 4.2). The inhalation toxicity reduced to milligrams per cubic metre per minute is 2220 mg/m^3/min LCL_0, or a LCt_{50} value of 2800 mg/m^3/min compared to 100 mg/m^3/min for nerve gas or 5000 mg/m^3/min for hydrogen cyanide. There is one piece of information relevant to humans. The first entry: 'ihl-hmn TCL_0 : 2 ppm' means that the lowest published toxic concentration of MIC when inhaled by humans is 2 ppm (4.7 mg/m^3). Toxic of course means any adverse reaction, not death. In summary, accepting all the dangers of extending data to humans, we can say that MIC is not particularly hazardous by mouth, but certainly has considerable inhalation toxicity, which explains the Bhopal tragedy. It is more toxic than hydrogen cyanide, which is itself a great hazard, especially when liberated in confined areas.

The book edited by Sax (5) gives a similar entry to that in RTECS, but has an expanded description of effects, to wit that MIC is highly irritant to skin, eyes and mucous membranes, and can cause pulmonary oedema (that is, increase in fluid in the lung). MIC is also stated to decompose rapidly in water.

We have enough information now to make an interpretation of the events of Bhopal. MIC was being stored for use to make carbaryl. Somehow it escaped, and the vapour attacked the lungs, eyes and skin of the sleeping victims. At a concentration of about 5 mg/m^3 it would produce irritation and some damage. This concentration would be unlikely to cause death, but a concentration of 50 mg/m^3 (21.5 ppm) persisting for 1 h ($50\ mg/m^3 \times 60\ min = 3000\ mg/m^3/min$) would be expected to produce many deaths. This is not a high concentration, particularly if tonnes are being released, for one tonne would contaminate 20 million m^3 of air to this level, or an area of nearly 0.2 km^2 with a vapour cloud height averaging 100 m. Death would be by asphyxiation caused by the lungs becoming full of fluid and ceasing to exchange oxygen and carbon dioxide. Many casualties would be expected; they would have varying degrees of lung, eye and skin injuries. The toxic effects reported for MIC are very like those of phosgene, which was an extremely lethal gas used in the First World War (Chapter 10). Because MIC decomposes rapidly in water, my guess is the effect is confined to the damp mucous membranes and it is the act of decomposition which is the toxic effect (as it is for phosgene). Therefore one would not expect that MIC would persist in the body. The toxic effect is exerted immediately on contact with the body tissues, and although much damage subsequently develops, this is all due to the short time of contact, not to the continuing presence of the toxic chemical.

By consulting a few secondary sources, we have been able to collect a lot of information which allows us to interpret an event in a chemical context. There is probably little point in going to primary sources of information. This practical example shows that it is not too difficult to use the technical literature.

The copy of RTECS that I consulted was on microfiche. I don't like microfiche, the reproduction is poor and sometimes unreadable, and I find that reading the magnified image on a screen is awkward. Further, it is very difficult to refer between pages of a document, i.e. to follow the main text and at the same time consult appendices or tables on other pages. I also mentioned the problems of storing paper hard copy, e.g. the 2.2 m of *Chemical Abstracts* from one year. Both of these two problem areas are likely to disappear in the near future due to the spread of 'computerised' information systems.

In fact, I could alternatively have consulted the Toxicology Data Bank, which has been introduced by the US National Library of Medicine. This source is an on-line computer file which gives much broader data for each entry than does RTECS but, at present, has considerably fewer entries.

It is a little inappropriate to talk of 'computerised' systems, for they do not compute in the mathematical sense. A better, but cumbersome, phrase is 'electronic data-storage and retrieval systems'. This emphasises the two aspects of storage, which is done on electromagnetic tape or disc, and sorting and retrieving, done by electronic logical circuitry which can recognise key words or other selected characters. Such systems have only come into general use since about 1980, but in another 5 years they may well have displaced many of the hard copy forms of secondary information, and also microfiche. The system differs from the present ones in that it is centralised. The data can be held in one 'bank', which may be on another continent. It is interrogated by means of telephone or cable links by the user, who has a keyboard to input request data, and a VDU (visual display unit) to observe the output from the data bank. This can also be printed at the user's terminal. There are various data banks, which may contain chemical, toxicological or regulatory data, and are compiled by staff of existing abstracting systems (e.g. *Chemical Abstracts*) or government authorities (e.g. National Institutes of Occupational Safety and Health, USA) or commercial organisations. The customer pays for the service on the basis of use, i.e. how long he is 'on-line' to the bank.

I do not propose to describe such systems in detail. Their use requires some training in the terminology of particular data banks and in the construction of a logical sorting system (search profile), so it is better to ask for assistance rather than try to interrogate the system oneself. Also, as the cost is determined by the time on-line; an inexperienced person may waste time and thus be a costly liability to whoever is getting billed for the service.

We may not see a complete displacement of hard copy abstracts by computer files, at least not in the immediate future. The hard copy is useful when planning a search; preliminary exploration of the printed material allows the construction of a much more effective and briefer computer search than is possible if you start directly on the computer. This is important in terms of cost, for if you start asking the computers lots of questions in the interactive mode, it becomes expensive (at about $2/min [Australian] in 1985). Further, it is often quicker to make a visual scan of an entry in hard copy and rapidly pick out data of interest than to have to ask several questions of the computer, and then not have a full comprehension of what data are available. The two systems are complementary; a preliminary hard copy study leads to a better computer search. In future, the design of the software for computer files may improve to the extent that they are more readily usable by the layman, and with increasing use costs may decline. Therefore hard copy in libraries will largely be restricted to the type of secondary information labelled A and D in Fig. 18.1, plus much of the primary information. Microfiche will largely disappear. The VDU, keyboard and printer (Fig. 18.2) will be standard items for any library and information body.

Interrogation of the Occupational Safety and Health (NIOSH) file for information on MIC gave eight entries. One entry had been duplicated, and was the *Archiv für Toxikologie* article that RTECS cited, which seems to be the only source of information on MIC toxicity. Of the remaining six entries, four referred to analytical techniques and two were discussions of the hazards associated with isocyanates in industry. The information base for the electronic system is the same as for hard copy. It is quicker to use by the experienced operator, is more easily updated and involves no storage at the customer's premises.

One may take a superficial view of this situation and say that it is very easy for you as a layperson to find out all you need to know about chemicals. Therefore you have no excuse for being misinformed or ignorant. This would be a judgement from a very privileged viewpoint, and certainly unfair in practice. The reason for this comment is that although the data are available, the value of these data really depends on how effectively they can be interpreted and put into the context of the situation which you are evaluating. Thus a knowledge of the LD_{50} value of MIC is of no use unless you are able, from previous knowledge, to rank it in an order with other industrial chemicals (Chapter 4). The symptoms of the victims can be

Fig. 18.2 The terminal to on-line computer files of data is now a standard feature of technical libraries. Shown here are the visual display unit, keyboard, printer and the information services librarian.

correlated with the published properties of MIC only by someone who has experience of how chemicals enter the body (Chapter 5) and attack the living tissues. There is a necessity for a background of experience, which will give practical meaning to the base data. In the absence of this it is possible to make glaring errors of interpretation. Therefore I do not think that the needs of society for information are well served in the broadest sense; we have a well-developed information technology which serves a few. What we need is better dissemination of that knowledge to the whole of society, free from partisan bias. I hope this book is a contribution to that end. Certainly you should seek for information, but I counsel care in its interpretation.

References

1. Weast, R. C. (ed., 1984). *CRC Handbook of Chemistry and Physics*. 65th edn. Boca Raton: CRC Press Inc.
2. Windholz, M. (ed., 1983). *The Merck Index*. 10th edn. Rahway: Merck & Co. Inc.
3. Brandrup, J. & Immergut, E. H. (eds, 1975). *Polymer Handbook*. 2nd edn. New York: Wiley-Interscience.
4. Lewis, R. J., Sr. & Sweet, D. V. (eds, 1984). *Registry of Toxic Effects of Chemical Substances*. Cincinnati: National Institute of Occupational Safety and Health.
5. Sax, N. I. (ed., 1984). *Dangerous Properties of Industrial Materials*. 6th edn. New York: Van Nostrand Reinhold Company.
6. Reynolds, J. E. F. (ed., 1982). *Martindale. The Extra Pharmacopoeia*. 28th edn. London: The Pharmaceutical Press.
7. Grayson, M. (ed., 1978–84). *Kirk-Othmer Encyclopedia of Chemical Technology*. 3rd edn. 24 vols plus supplement and index. New York: Wiley-Interscience.

19

Summary and conclusions

I will briefly summarise what we have discussed so far, then attempt to make some general conclusions. Inevitably this latter process will reflect my own opinions; you will have to test their quality against the facts that we have established. Lastly, we can try to assess how chemicals will affect our society in the future.

Over 40 years, synthetic organic chemicals have been developed in variety and quantity of production to the extent that they intrude on all areas of our society. They have brought material benefits, which must however be judged by broad social values which are not easy to assess. Among the problems which have arisen are those of toxic chemicals. The concept of toxicity must be understood before we can evaluate this situation, and it is clear that all chemicals are toxic. Therefore we must know the dose of a chemical necessary to produce a toxic effect, and how that chemical may enter the human body. We must consider very carefully all the available information on the toxicity of a chemical before we make statements about its likely effect. Then we also have to consider the real situation in which people might be exposed to the chemical, in order to assess whether they will be at risk. Thus even the most toxic chemical is not a risk if the quantity that could contact a person is below a toxic dose.

Talk of quantity leads to the involved topic of chemical analysis. This again has pitfalls to trap the unwary enthusiast who rushes forward down the trail of preconceived notions. Both toxicity and analysis are concepts with a strong comparative nature; absolutes are foreign to them. We need to know what is in a sample, and secondly whether the concentration found is significant or trivial. In analysis, as in toxicity, we must make relative judgements based on measurements; we cannot say: 'It is there, it is toxic, therefore it is a hazard'.

Finding out what are hazards is not easy either. We are exposed to all sorts of hazards which have nothing to do with chemicals, many of which have threatened our race for millennia. It is not easy to find new problems which might be due to chemicals among all the other problems and hazards. When we have established some estimate of the hazard associated with a chemical, the next step is to compare that hazard with the benefits to be gained from the use of the chemical. If the benefits are small, e.g. the chemical is used as a cosmetic, then its use can be stopped. If there are very real benefits, then we may consider it worthwhile to continue using a chemical which also is hazardous in use (e.g. a pesticide).

A frightening hazard is that of the development of cancer. However, there is no evidence to indicate that the New Chemical Age has initiated an epidemic of cancers, although some few chemicals are known to have caused cancers in workplaces. By contrast, it is not in doubt that a rise in the incidence of lung cancer has been caused by cigarette smoking. Certainly we must continue to monitor the incidence of cancer, particularly in workplaces where exposure to chemicals is great, and quicker methods for predicting whether a chemical is a carcinogen must be sought.

Over the past 200 years the use of chemicals as medicinal drugs has been rationalised, and the drugs are now made by synthetic methods which ensure purity and potency. The present situation might be wholly admirable, if it were not for the human failings which have caused many people to reject orthodox medicine and return to the obscure practices of the past. What should therefore have been one of the great chemical success stories, the bright example of benefit to humans rendered by the Chemical Age is, in many persons minds, a tarnished and ambiguous victory. We can see deficiencies both in the current practice of orthodox medicine and in the logic of its opponents. The success in a technical sense is clear; we have drugs which control many diseases. It is our human inability to administer and control them which has muddied the situation. There is also an element of obscurantism which haunts several of the topics we are considering.

The use of chemicals in warfare was intense during 1915–18, but by some miracle we escaped a chemical war in the 1939–45 conflict. In 1980 we might confidently have expected that a chemical disarmament treaty or convention was probable, and would remove at least one horror from warfare. Now (in 1985) the situation is much less hopeful. The nations that profess great interest in a convention upon the public stage in Geneva show no intention at all of actually doing anything about a real chemical war in the Persian Gulf, nor of condemning a clear breach of the 1925 Geneva Convention by Iraq. The two superpowers are trying only to make political capital of two supposed chemical warfare events in southeast Asia (herbicides in Vietnam versus Yellow Rain) whereas many other leading countries have chemical industries whose exports are sacrosanct. Until nations can put principles before profits, we will not see any diminution in the risk of chemical war nor, for that matter, of any other form of war.

If we do not all go out rapidly in a chemical war, we have to counter the possibility of us all being slowly poisoned by an accumulation of chemical junk in the environment. I do not believe there is much risk of this, provided we put effort and finance into developing the technical means to control pollution. Because of public concern and agitation about pollution, we are much more alert than 10 years ago, and it is unlikely that the gross abuses of the past will be allowed to recur. However, we must ensure that the present efforts continue, that there is not a swing against pollution-control measures. My main fear is that we may put great effort into finding solutions for specific small parts of the problem simply because they are ones which have attracted public attention and, by so doing, neglect to consider the full range of the pollution question. In other words, devoting all effort to PCBs and dioxins may clean up relatively minor nuisances, but what we need is a broad strategy which takes account of the bulk of slightly toxic chemical products. Here is a test of our rationality: we have, or can develop, the technical tools to control pollution; if we fail it would be entirely our own fault. Failure must imply the inability to use our techniques to a rational end.

We need to protect ourselves against chemicals, mainly in the workplace, hopefully not in the environment. Much technical information is obtainable and we have equipment to deal with any level of chemical threat. The human attitudes towards protective equipment are much more of a complex question. Provision of protective gear is a relatively easy matter, but to ensure that it is worn correctly is not. It is desirable that technical improvements continue to be sought with the main aim to make the gear less burdensome to wear. We also need to do much more training in the use and wearing of the equipment, so that the user is familiar with it and values it. Merely handing out respirators or gloves or protective suits is pointless; workers must want to wear them because they understand their value.

The discussion about the herbicides of the 2,4-D/2,4,5-T type (Chapters 13 and 14) is an example for you to peruse. We do not need to draw any conclusions from it, save to realise the complexity that must arise if we consider any one issue in depth. The two chapters are a severe condensation of the issue anyway; I will, no doubt, be accused of omitting vital aspects of the case.

There will be no control of chemicals unless we legislate for it, and then enforce the consequent regulations. When control costs money, regulation alone will ensure that it is applied. To put such legislation on a rational footing requires much technical knowledge, and it is clear that at present much of this is lacking. Regulation also brings the chemist into contact with that strange and capricious beast, the law. At least, that is how chemists view it, and I suspect lawyers see chemists in a similar light. In other words, we have to educate the lawyers in the basic concepts of chemically related topics, principally toxicity and analysis. I mentioned in Chapter 6 the Victorian law concerning zero alcohol levels for probationary drivers. When the interpretation of the law was questioned during its passage through Parliament, the comment was, let the courts work out an interpretation. This is an absurd shirking of responsibility, and can only result in regulation being based on pseudoscience; the lawyers' interpretation of a technical issue. We must ensure that all legislation concerning chemicals is based on sound chemistry; unambiguous to the courts and not in need of interpretation.

The notion that the natural world of chemicals that we left 40 years ago was totally beneficient is a false one, as is the related concept that synthetic chemicals are all nasty. Natural food can carry natural poisons; natural drugs can have side-effects, as synthetic ones do. There is no division or distinction between natural and synthetic chemicals, save in their origin. Further, we cannot displace chemical science with religion. Science and religion can coexist as complementary aspects of the human mind, but rational argument cannot be dismissed by dogma.

Information is available to us from sources which are reasonably accessible but, to obtain it, we must first have the will to get it and then we have to understand the basics of modern information systems. When the information has been outlined, then it may require a degree of interpretation, so that the public needs the help of guides to find the chemical facts, and assistance from chemists in interpreting them. However, I believe you should acquire as much information as you can by your own efforts, which will free you from the subtle (or blatant) snares of those who want to convert you to a particular belief. Rational judgement must be founded on knowledge; the provision of the latter is a necessity for a rational appraisal of the chemical world.

In fact, if we attempt to draw general conclusions from the above summary, I think a major one is that we are seeing a departure from rationalism in a number of aspects of the Chemical Age. I referred to the hope that was current in 1945 which saw science being put to rational use to improve the world. Now that hope is faded or, more exactly, it has been reversed into a fear that science will bring catastrophe of a chemical, nuclear or other type. In our discussions on drugs, foods, herbicides and many other issues this tendency to go from science to the obscure has been evident. It is seen in areas other than chemistry; the growth of creationism is a puzzling example. Somehow science is not enough for people or, at least, has not been perceived as being sufficient. To a large degree this situation must have arisen from a failure to communicate between the scientist and the public.

I have discussed aspects of the failure to communicate in various parts of this book, principally in Chapters 1 and 14. The media need to improve their presentation of chemical science to the public. I believe this can be done in an acceptable manner without turning the public away, or the television set off. More journalists and reporters should have a technical background and be encouraged to research topics. What annoys me is the incredible laziness of reporters when it comes to researching a technical matter. All they do is to find the nearest 'expert' (credentials are unimportant) and shove a microphone in that person's face. All sorts of interesting information may be lying about in the public domain, but if this is not repeated to the journalist in words of one syllable, the industry/government department/research laboratory is accused of keeping it secret. The technicalities of the presentation to the public are often good, what is poor is content. This is the result of the inability to pick the appropriate expert. The people most willing to talk are usually those with a cause to push, or a crusading desire to push some crank notion. Those with the best knowledge of a chemical topic are either not contacted or unwilling to be drawn into a crazy argument; they are also aware that the reporter will not scruple to distort or mutilate their comments if that action will sensationalise a dull factual account. The unfortunate reporter thus has to rely on whoever is willing to talk; garrulity and veracity are not close companions. This situation is self-reinforcing. Chemist does not talk to reporter because the media always present oddball views. Reporter depends on self-appointed experts because the knowledgeable excuse themselves from interviews.

It follows from the above that much discussion about the chemical world is heated discussion between factions of extreme views, neither of which may be impeded by facts in presenting their arguments. Self-interest is startlingly effective in altering your perceptions about facts, so that if you are protecting a profitable industry your arguments may not be objective. Equally subjective arguments may be advanced by those who have beliefs founded on irrational premises. The mass of the public is caught in the middle, battered by the surging to and fro of the extremists. However, I do not believe that we, as the public in the middle, should passively wait to be battered. In many ways it is easy to adopt extreme views for, after one act of faith, there is no further need for rational effort. Accept the dogma and follow the leader. In the middleground of society, every topic must be investigated and assessed before an opinion is formed. Thus only the industrious continue to hold firm, the lazy move to the extremes and grab a banner. If society is to continue on a rational basis, it is imperative that we look to the defence of the middleground. It cannot be held by passive methods but must be defended energetically by knowledge and reason. This applies to chemical matters, to political matters, to all topics that concern society. Lose the middleground and the Dark Ages return.

What, then, have chemists done to defend their middleground? The learned societies and professional bodies have to an extent endeavoured to explain chemically related matters to the public, but they are largely inwardlooking; chemist talking to chemist rather than to the public. I confess that I personally have done little until now to assist the public. It seems that chemists should strive much more vigorously to present their professional image to the public and to offer their services as sources of chemical information. I cannot think of any chemical, biochemical or toxicological society that issues news releases on topics of current interest. If they do, the press must totally ignore them. Contrast this situation with that of lawyers or medical practitioners. Lawyers have advanced the ambit of their operations well beyond the strict practice of law so that it embraces many aspects of public life. In Australia this has resulted in lawyers making decisions on the economy and on technical matters. The fixing of wages before the Arbitration Commission is conducted as though it were a law suit, with counsel on each side and a judge presiding. If there are technical matters of national impact to be investigated, set up a Royal Commission with another pair of legal teams to settle the matter. It is of no concern to anyone that the legal system of argument between professional antagonists in court is not a good way to decide scientific matters. Nor, apparently, is it of any concern that the lawyers very frequently do not understand the issues they are arguing about. Certainly they will not object; their fees are substantial.

Doctors are frequently regarded as oracles on matters other than the cure of disease; a doctor's pronouncement will gain more credence than that of a scientist. However, doctors are not trained in rational deduction or analytical thinking, they are trained to respond quickly to a situation in order to save life or mitigate suffering. Their decisions have to be made rapidly on limited evidence. This is perfectly proper training, for the patient is unlikely to be grateful if he dies this week whilst waiting for an accurate diagnosis available next week. I am not belittling the medical profession, it does a very practical job in circumstances which are often trying. I am merely pointing out that its members are not theorists and are not infallible connectors of effect with cause.

There is a need for professional chemical groups to assume a proper public role, if necessary elbowing the lawyers back to law and standing up to the medical profession on matters of chemical science. If this action is taken, the public awareness of chemically related matters should improve greatly.

What of the future? There will be more Bhopals, thalidomides and Times Beaches, though we may hope that some lessons have been learnt. Such incidents are minor when considered over the whole range of human activity. I think, therefore, that the main problem is the question of chemical disposal, the broad problem of the mass of material of low toxicity which nevertheless can slowly gum up our environment. We have the technical knowledge to solve many problems and, if this is coupled to a social will, then these problems will disappear. This presupposes that we can reverse the obscurantist tendencies of recent years.

The future use of chemicals as drugs, food additives and pesticides will also depend on a rational appreciation of advantages and disadvantages. We may totally reject their use and thus renounce all science and technology in a fit of obscurantism. Alternatively we can let our chemical industries go full bore, pouring out limitless supplies of chemicals for the best profits in the best of all chemical worlds. We will in fact finish somewhere between these extremes, but where will depend on our awareness of problems. We keep coming back to this question of knowledge and information. Perhaps this is the one simple answer to our chemical problems: good diffusion of technical information to the public will ensure a safe chemical future.

So we have the means to control the future if we are wise. I hope this book is a useful tool in the spread of knowledge about the New Chemical Age. Remember one thing: I do not give a damn about the beliefs that you finally sustain concerning chemicals. All I ask is that those beliefs have a rational basis.

Index

Page numbers in *italics* refer to tables and figures.